21世纪项目案例开发规划丛书

# Android App 贯穿式 项目实战

微课视频版

兰红 王建雄 井福荣 ◎ 编著

清华大学出版社
北京

## 内 容 简 介

本书是根据国家一流专业建设教学改革的需要,结合作者讲授"移动应用开发"课程的教学经验编写而成的。全书共 12 章,以一个 Android+Java Spring 的完整项目案例,贯穿了书中的主要知识点;同时在每个章节提供了独立的案例帮助读者理解内容和提升编程能力。前 5 章主要包括 Android 界面设计和事件处理机制,学习完前 5 章读者可以实现基本的 App 设计。第 6~8 章主要介绍 Android 的数据存储形式、数据操作和后台服务,学习完前 8 章读者可以完成功能较完备的单机版 App 开发。第 9~11 章主要介绍网络版开发,即目前主流的 Android 前端+后台服务端开发模式,首先介绍 Android 的网络通信编程,之后介绍 Android+PHP 的环境配置和开发,并给出了"倾心家教"项目开发案例。第 12 章介绍基于 Java Spring Boot 框架开发后台的"橙一色"App 案例,为掌握前后端相结合的项目开发起到抛砖引玉的作用。本书内容丰富,实用性强,侧重案例教学和计算机程序设计的基本知识,帮助读者掌握 Android 的核心内容及 Android App+服务器端设计的基本方法和编程技巧,还针对知识点提供了微课视频讲解。

本书可作为高等学校相关课程的教学用书,也可作为 Android 应用的社会培训教材及计算机爱好者的参考书籍。

本书封面贴有清华大学出版社防伪标签,无标签者不得销售。
版权所有,侵权必究。举报: 010-62782989, beiqinquan@tup.tsinghua.edu.cn。

**图书在版编目(CIP)数据**

Android App 贯穿式项目实战:微课视频版/兰红,王建雄,井福荣编著. —北京:清华大学出版社,2020.11 (2024.1 重印)
(21 世纪项目案例开发规划丛书)
ISBN 978-7-302-55785-2

Ⅰ. ①A… Ⅱ. ①兰… ②王… ③井… Ⅲ. ①移动终端-应用程序-程序设计 Ⅳ. ①TN929.53

中国版本图书馆 CIP 数据核字(2020)第 105284 号

策划编辑:魏江江
责任编辑:王冰飞　张爱华
封面设计:刘　键
责任校对:李建庄
责任印制:丛怀宇

出版发行:清华大学出版社
网　　址:https://www.tup.com.cn, https://www.wqxuetang.com
地　　址:北京清华大学学研大厦 A 座　　邮　编:100084
社 总 机:010-83470000 　　邮　购:010-62786544
投稿与读者服务:010-62776969, c-service@tup.tsinghua.edu.cn
质量反馈:010-62772015, zhiliang@tup.tsinghua.edu.cn
课件下载:https://www.tup.com.cn, 010-83470236

印 装 者:三河市龙大印装有限公司
经　　销:全国新华书店
开　　本:203mm×260mm　　印　张:23.25　　字　数:608 千字
版　　次:2020 年 12 月第 1 版　　印　次:2024 年 1 月第 4 次印刷
印　　数:6001~7000
定　　价:89.90 元

产品编号:088015-01

# 前言

## Android 的发展趋势

Android（安卓）是一种基于 Linux 的自由及开放源代码的操作系统，由 Google 公司和开放手机联盟领导及开发，主要用于移动终端设备，如市场上的智能手机和平板电脑。Android 系统平台以开源性和丰富的可扩展性受到用户好评，国内拥有数量庞大的智能 Android 手机用户群，手机管理软件凭借丰富的 App 应用资源下载和便捷的管理功能，成为 Android 手机用户的装机必备选择。

大量的用户需求使得 Android App 开发仍然以源源不断的上线方式来展现。从普通大众的消费水平以及使用习惯上看，Android App 开发的市场还是很广阔的。不少游戏平台都转向 Android 手机，对 Android 游戏 App 开发将会持续增多。放眼应用市场，不难发现 Android App 开发所涵盖的类型和领域非常多，游戏、社交、旅游、工具等类型的应用都有大量的 Android 系统开发。Android 开发产品的数量会不断增加，质量也会不断改进。

## 本书的编写安排

本书可以作为 Android 开发的一本入门图书，通过理论知识与大量的案例来介绍 Android 应用开发的各方面知识。在学习本书之前，需要读者具备 Java 基础知识，因为 Android 开发使用的是 Java 语言，建议读者先了解理论知识，掌握组件的使用方式，然后通过具体的案例来达到熟练应用。

本书共分为 12 章，具体如下：

第 1 章主要介绍 Android 的基础知识，包括 Android 的发展史、Android 的系统架构、开发环境的搭建、第一个 Android 项目的创建、项目的文件结构。通过这些基础知识让开发者对 Android 项目的创建及目录有一个简单的了解。

第 2、3 章主要介绍 Android 的布局以及 Activity，包括 Android 的六大布局、各种组件的使用、AdapterView 及其子类的使用、Intent 的使用方式。通过这部分讲解可以让开发者实现简单的用户注册。

第 4、5 章主要介绍 Android 的事件处理机制和 Fragment，讲述 Android 事件处理机制的方式、异步类的使用、Fragment 的生命周期以及 Fragment 与 Activity 之间的通信。

第 6～8 章主要介绍 Android 的数据存储、内容提供者以及服务和广播的使用。在这几章中，针对每个知识点都通过具体的案例来讲解，让开发者快速地掌握 Android 开发的几大组件。

第 9、10 章主要介绍 Android 的网络通信编程，包括 HTTP 通信、Socket 通信、数据提交方式以及Android+PHP开发。这两章通过从网络下载图片在应用程序中的显示来讲解HTTP通信,通过搭建本地PHP开发环境来讲解Android和本地服务器的通信,让开发者对Android的网络编程有基本的了解。

第11章主要通过具体的案例("倾心家教"应用案例)开发来讲解Android+PHP+MySQL的使用。本章从项目的需求分析、界面设计、数据库设计、功能实现来完整地讲解Android项目的开发流程。

第12章主要通过具体的案例("橙一色"App项目案例)开发来讲解Android+Spring Boot+MySQL的使用。本章从项目的需求分析、界面设计、数据库设计、功能实现来完整地讲解Android前端+服务器端项目的完整开发流程。

## 本书的配套资源

本书提供教学大纲、教学课件、电子教案、习题答案、程序源码和教学进度表等配套资源,本书还提供900分钟的微课视频。

---
**资源下载提示**

课件等资源:扫描封底的"课件下载"二维码,在公众号"书圈"下载。

素材(源码)等资源:扫描目录上方的二维码下载。

视频等资源:扫描封底刮刮卡中的二维码,再扫描书里章节中的二维码,可以在线学习。

---

## 致谢

本书的编写由兰红、王建雄、井福荣共同完成。特别感谢李志军、朱合隆、方治屿同学,"倾心家教"和"橙一色"App项目案例均为大学生创新项目作品,目前已投入使用。感谢徐民霖、李浩瀚、王坤、朱纯煜等同学对文稿的校对,衷心感谢支持本书出版的各位领导和同事,感谢为本书顺利出版做出努力的清华大学出版社。

## 意见反馈

本书代码基于Android 8.0版本,在Android Studio上验证通过。由于水平有限,书中难免存在疏漏之处,请广大读者批评指正。

作 者
2020年7月

# 目录
## CONTENTS

源码下载

第 1 章　Android 入门 ································································ 1

  1.1　Android 概述 ······························································· 1
    1.1.1　无线通信技术 ······················································ 1
    1.1.2　Android 基本介绍 ················································ 2
    1.1.3　Android 系统架构 ················································ 3
  1.2　搭建 Android 开发环境 ··············································· 4
  1.3　开发第一个 Android 程序 🎬 ······································ 6
    1.3.1　创建与运行第一个 HelloWorld 程序 ······················ 6
    1.3.2　学习项目文件 ······················································ 8
    1.3.3　Android 项目结构 ················································ 9
    1.3.4　AndroidManifest.xml 详解 ····································· 9
  1.4　【项目案例 1】"橙一色"App 项目创建 🎬 ················ 10
  1.5　本章小结 ······································································ 11
  1.6　课后习题 ······································································ 12

第 2 章　Android 应用界面 ······························································ 13

  2.1　View 概念 ···································································· 13
  2.2　布局管理器 🎬 ···························································· 15
    2.2.1　创建和使用布局文件 ·········································· 15
    2.2.2　布局的类型 ························································ 16
  2.3　Android 控件详解 🎬 ················································· 23
    2.3.1　TextView ······························································ 23
    2.3.2　EditText ································································ 24
    2.3.3　Button ·································································· 25
    2.3.4　ImageView ···························································· 26
    2.3.5　RadioButton 和 CheckBox ····································· 27

|  |  |  |
|---|---|---|
| | 2.3.6 ProgressBar | 28 |
| | 2.3.7 SeekBar | 30 |
| 2.4 | AdapterView 及其子类 | 31 |
| | 2.4.1 ListView 和 ListActivity | 31 |
| | 2.4.2 Adapter 接口 | 33 |
| 2.5 | 对话框的使用 | 40 |
| 2.6 | Toast 的使用 | 42 |
| 2.7 | Log 的使用 | 43 |
| 2.8 | 用户注册案例布局讲解 | 44 |
| 2.9 | 【项目案例 2】"橙一色"App 注册/登录界面设计 | 47 |
| 2.10 | 本章小结 | 55 |
| 2.11 | 课后习题 | 55 |

第 3 章　Activity ································································ 56

　　3.1　Activity 基础 ································································ 56
　　　　3.1.1　认识 Activity ································································ 56
　　　　3.1.2　如何创建 Activity ································································ 57
　　　　3.1.3　Activity 的生命周期 ································································ 58
　　　　3.1.4　Activity 中的单击事件 ································································ 63
　　3.2　Intent 的使用 ································································ 64
　　　　3.2.1　Intent 浅析 ································································ 64
　　　　3.2.2　显式 Intent ································································ 64
　　　　3.2.3　隐式 Intent ································································ 65
　　3.3　Activity 中的数据传递方式 ································································ 66
　　3.4　用户注册案例功能实现 ································································ 67
　　3.5　本章小结 ································································ 70
　　3.6　课后习题 ································································ 70

第 4 章　Android 事件处理 ································································ 71

　　4.1　Android 事件处理机制 ································································ 71
　　4.2　基于监听的事件处理 ································································ 71
　　4.3　基于回调的事件处理 ································································ 73
　　4.4　AnsyncTask 异步类的功能与用法 ································································ 75
　　4.5　【项目案例 3】"橙一色"App 注册验证 ································································ 78
　　4.6　本章小结 ································································ 83
　　4.7　课后习题 ································································ 83

第 5 章　Fragment 基础 ································································ 84

　　5.1　Fragment 概述 ································································ 84

5.2 Fragment 的生命周期 ············································································· 85
5.3 Fragment 的创建 ················································································· 86
5.4 Fragment 与 Activity 之间的通信 ····························································· 91
5.5 【项目案例 4】"橙一色"App 商品展示功能实现 ········································ 97
5.6 本章小结 ···························································································· 103
5.7 课后习题 ···························································································· 103

## 第 6 章 Android 数据存储 ·············································································· 104

6.1 数据存储方式简介 ················································································ 104
6.2 文件存储 ···························································································· 105
    6.2.1 外部存储 ··················································································· 105
    6.2.2 内部存储保存用户注册信息 ·························································· 107
6.3 XML 文件的序列化与解析 ····································································· 111
    6.3.1 XML 序列化 ············································································· 111
    6.3.2 XML 序列化实例 ········································································ 112
    6.3.3 XML 文件解析 ··········································································· 116
    6.3.4 XML 解析实例 ··········································································· 116
6.4 SharedPreferences 的使用 ······································································· 121
    6.4.1 SharedPreferences 简介 ································································· 121
    6.4.2 使用 SharedPreferences 存储用户注册信息 ······································· 122
6.5 SQLite 数据库 ····················································································· 127
    6.5.1 SQLite 数据库简介 ····································································· 127
    6.5.2 SQLite 数据库操作类以及接口 ······················································ 127
    6.5.3 SQLite 数据库的操作 ·································································· 129
    6.5.4 使用 SQLite 数据库展示用户信息 ·················································· 132
6.6 JSON 学习 ························································································· 143
    6.6.1 JSON 简介 ················································································ 143
    6.6.2 JSON 解析案例 ·········································································· 145
6.7 【项目案例 5】"橙一色"App 数据存储及主页展示 ···································· 149
6.8 本章小结 ···························································································· 161
6.9 课后习题 ···························································································· 162

## 第 7 章 ContentProvider 的使用 ······································································ 163

7.1 ContentProvider 简介 ············································································ 163
7.2 操作 ContentProvider ············································································ 164
    7.2.1 ContentProvider 的创建 ································································ 164
    7.2.2 Uri 简介 ···················································································· 165
    7.2.3 使用 ContentProvider 获取通讯录 ··················································· 165

7.3 使用 ContentProvider 共享数据 … 168
    7.3.1 ContentResolver 简介 … 168
    7.3.2 系统短信备份案例 … 169
    7.3.3 跨应用数据共享案例 … 175
7.4 ContentObserver … 181
    7.4.1 ContentObserver 简介 … 181
    7.4.2 监控短信发送案例 … 182
7.5 本章小结 … 186
7.6 课后习题 … 186

## 第 8 章 Service 和广播的使用 … 187

8.1 Service 简介 … 187
    8.1.1 Service 的创建和配置 … 187
    8.1.2 Service 的启动与停止 … 188
8.2 Service 的生命周期 … 195
8.3 Service 通信 … 196
    8.3.1 本地服务和远程服务通信 … 196
    8.3.2 本地服务通信实例 … 198
    8.3.3 跨应用启动服务实例 … 205
8.4 系统服务类的使用 … 212
    8.4.1 TelephonyManager … 212
    8.4.2 SmsManager … 214
8.5 广播消息 … 216
    8.5.1 广播简介 … 216
    8.5.2 静态注册广播案例 … 217
    8.5.3 动态注册广播案例 … 219
8.6 【项目案例 6】"橙一色"App 消息推送功能实现 … 222
8.7 本章小结 … 225
8.8 课后习题 … 225

## 第 9 章 Android 网络通信编程 … 226

9.1 网络编程基础 … 226
    9.1.1 HTTP 简介 … 226
    9.1.2 标准 Java 接口 … 227
    9.1.3 Android 网络接口 … 228
9.2 HTTP 通信 … 228
    9.2.1 HttpURLConnection 简介 … 228
    9.2.2 HttpURLConnection 接口使用案例 … 230

9.2.3　HttpClient 简介 ……………………………………………………… 233
9.2.4　HttpClient 的使用案例 ………………………………………………… 234
9.3　Socket 通信 ……………………………………………………………………… 237
9.3.1　Socket 通信原理 ………………………………………………………… 237
9.3.2　Socket 通信案例 ………………………………………………………… 239
9.4　数据提交方式 ……………………………………………………………………… 242
9.4.1　GET 方式提交数据 ……………………………………………………… 242
9.4.2　POST 方式提交数据 …………………………………………………… 243
9.5　OkHttp 的使用 …………………………………………………………………… 244
9.5.1　OkHttp 的 GET 和 POST 请求 ………………………………………… 244
9.5.2　在 Android 中使用 OkHttp …………………………………………… 246
9.6　本章小结 …………………………………………………………………………… 248
9.7　课后习题 …………………………………………………………………………… 248

## 第 10 章　Android＋PHP 开发 …………………………………………………………… 249

10.1　PHP 介绍 ………………………………………………………………………… 249
10.2　PHP＋MySQL 简介 …………………………………………………………… 252
10.2.1　PHP 连接 MySQL ……………………………………………………… 252
10.2.2　PHP 创建数据库 ……………………………………………………… 253
10.2.3　PHP 创建数据表 ……………………………………………………… 255
10.2.4　PHP 对数据库表的基本操作 ………………………………………… 256
10.3　PHP＋Android 简介 …………………………………………………………… 258
10.3.1　Android 与 PHP 结合 ………………………………………………… 258
10.3.2　用户登录案例 ………………………………………………………… 261
10.4　本章小结 ………………………………………………………………………… 265
10.5　课后习题 ………………………………………………………………………… 266

## 第 11 章　"倾心家教"应用案例开发 …………………………………………………… 267

11.1　应用分析 ………………………………………………………………………… 267
11.2　应用界面设计 …………………………………………………………………… 268
11.2.1　登录界面设计 ………………………………………………………… 268
11.2.2　主界面规划设计 ……………………………………………………… 271
11.2.3　找教师界面 …………………………………………………………… 276
11.2.4　消息界面 ……………………………………………………………… 281
11.2.5　个人信息界面 ………………………………………………………… 286
11.3　数据库设计 ……………………………………………………………………… 288
11.4　应用功能实现 …………………………………………………………………… 291
11.4.1　应用主界面实现 ……………………………………………………… 291

11.4.2　教师查询/预约功能实现 …………………………………………… 300
　　11.4.3　个人信息管理功能实现 …………………………………………… 308
　　11.4.4　预约的教师查询功能 ……………………………………………… 314
　　11.4.5　设置功能的实现 …………………………………………………… 317
11.5　应用发布 …………………………………………………………………… 318
11.6　本章小结 …………………………………………………………………… 320

## 第12章　"橙一色"App 功能优化及完善 …………………………………………… 321

12.1　应用分析 …………………………………………………………………… 321
12.2　App 应用技术 ……………………………………………………………… 321
　　12.2.1　Spring Boot 框架 …………………………………………………… 321
　　12.2.2　MyBatis 持久层框架 ………………………………………………… 325
　　12.2.3　Fastjson 解析库 ……………………………………………………… 327
12.3　"橙一色"App 应用功能界面设计 ………………………………………… 330
　　12.3.1　商品界面 ……………………………………………………………… 330
　　12.3.2　购物车界面 …………………………………………………………… 331
12.4　数据库创建 ………………………………………………………………… 333
12.5　Web 前端界面设计 ………………………………………………………… 334
12.6　Spring Boot 服务端设计 …………………………………………………… 336
　　12.6.1　用户接口的设计 ……………………………………………………… 336
　　12.6.2　商品接口的设计 ……………………………………………………… 339
　　12.6.3　消息接口的设计 ……………………………………………………… 343
　　12.6.4　销量接口的设计 ……………………………………………………… 344
12.7　"橙一色"App 端功能实现 ………………………………………………… 345
　　12.7.1　App 登录/注册实现 ………………………………………………… 345
　　12.7.2　App 首页商品信息展示功能实现 ………………………………… 348
　　12.7.3　App 商品信息展示功能实现 ……………………………………… 351
　　12.7.4　App 消息推送功能实现 …………………………………………… 353
12.8　本章小结 …………………………………………………………………… 357

**参考文献** ………………………………………………………………………………… 358

# 第 1 章

# Android入门

**学习目标**
- 了解 1G、2G、3G、4G、5G 无线通信技术。
- 掌握开发环境的搭建。
- 动手开发第一个 Android(安卓)HelloWorld 程序。

Android 是基于 Linux 开放性内核的操作系统,是 Google 公司在 2007 年 11 月 5 日公布的手机操作系统。自问世以来,Android 就受到了众多关注,并成为移动平台最受欢迎的操作系统之一。本章将针对 Android 的入门知识进行详细的讲解。

## 1.1 Android 概述

### 1.1.1 无线通信技术

在学习 Android 系统之前需要先了解通信技术方面的知识。随着智能手机的发展,通信技术也从最开始的 1G、2G 发展到 3G、4G,现在 5G 技术也已经来到,接下来将详细地讲解这 5 种技术。

- 1G。1G 的移动通信电话用的是模拟蜂窝通信技术,这种技术只能提供区域性语音业务,而且通话效果差,保密性能也不好,用户的接听范围也很有限。
- 2G。2G 指第二代通信技术,2G 技术分为窄带 TDMA(Time Division Multiple Access,时分多址)、GSM(Global System for Mobile Communications,全球移动通信系统)和 CDMA(码分多址)共 3 种。TDMA 是欧洲标准,允许在一个射频同时进行 8 组通话。GSM 具有较强的保密性和抗干扰性、音质清晰、通话稳定等优点。CDMA 技术完全适应现代移动通信网所要求的大容量、高质量、综合业务等。
- 3G。3G 是 3rd-Generation 的简称,是无线通信与互联网结合的移动通信系统,如视频聊天、语音聊天、在线购物、网游等。3G 技术在传输声音和数据的速度上有很大的提升。
- 4G。LTE(Long Term Evolution,长期演进技术)是 3G 的演进。就在 3G 通信技术正处在

酝酿之中时,更高的技术应用已经在实验室进行研发。4G 通信提供了一个比 3G 通信更完美的无线世界,它可以创造出许多消费者难以想象的应用。4G 手机可以提供高性能的流媒体内容,并通过 ID 应用程序成为个人身份鉴定设备。

- 5G。5G 是最新一代蜂窝移动通信技术,也是继 4G、3G 系统之后的延伸。5G 的性能目标是高数据速率、减少延迟、节省能源、降低成本、提高系统容量和大规模设备连接。5G 网络的数据传输速率远远高于以前的蜂窝网络,最高可达 10Gb/s,比当前的有线互联网要快,比先前的 4G LTE 蜂窝网络快 100 倍。它的另一个优点是较低的网络延迟(更快的响应时间),低于 1ms,而 4G 为 30~70ms。由于数据传输更快,5G 网络将不仅仅为手机提供服务,而且还将成为一般性的家庭和办公网络提供商,与有线网络提供商竞争。

### 1.1.2 Android 基本介绍

Android 一词的本义指"机器人",同时也是 Google 公司于 2007 年 11 月 5 日宣布的基于 Linux 平台的开源手机操作系统的名称,该平台由操作系统、中间件、用户界面和应用软件组成。

Android 一词最早出现于法国作家利尔亚当在 1886 年发表的科幻小说《未来夏娃》中。他将外表像人的机器起名为 Android。

Android 的 Logo 是由 Ascender 公司设计的,诞生于 2010 年,其设计灵感源于男女厕所门上的图形符号,于是伊琳娜·布洛克(Erina Blok)绘制了一个简单的机器人,它的躯干就像锡罐的形状,头上还有两根天线,Android 小机器人便诞生了,其中的文字使用了 Ascender 公司专门制作的称为 Droid 的字体。Android 是一个全身绿色的机器人,绿色也是 Android 的标志。颜色采用了 PMS376C 和 RGB 中十六进制的 ♯A4C639 来绘制,这是 Android 操作系统的品牌象征。有时候,它们还会使用纯文字的 Logo。Android 图标如图 1-1 所示。

图 1-1 Android 图标

2012 年 7 月美国科技博客网站 BusinessInsider 评选出 21 世纪十款最重要的电子产品,Android 操作系统和 iPhone 等榜上有名。

Android 系统版本如下。

- Android 1.1:发布时间为 2008 年 9 月。
- Android 1.5:Cupcake(纸杯蛋糕),发布时间为 2009 年 4 月。
- Android 1.6:Donut(甜甜圈),发布时间为 2009 年 9 月。
- Android 2.0/2.1:Eclair(松饼),发布时间为 2009 年 10 月 26 日。
- Android 2.2:Froyo(冻酸奶),发布时间为 2010 年 5 月 20 日。
- Android 3.0:Honeycomb(蜂巢),发布时间为 2011 年 2 月 20 日。
- Android 4.0:Ice Cream SandWich(冰激凌三明治),发布时间为 2011 年 10 月。
- Android 4.4:KitKat(奇巧),发布时间为 2013 年 9 月 4 日,API 等级为 19~20。

- Android 5.0-5.1.1：Lollipop（棒棒糖），发布时间为 2014 年 10 月 15 日，API 等级为 21～22。
- Android 6.0-6.0.1：Marshmallow（棉花糖），发布时间为 2015 年 5 月 28 日，API 等级为 23。
- Android 7.0-7.1.2：Nougat（牛轧糖），发布时间为 2016 年 8 月 22 日，API 等级为 24～25。
- Android 8.0-8.1：Oreo（奥利奥），发布时间为 2017 年 8 月 22 日，API 等级为 26～27。
- Android 9.0：Pie（馅饼），发布时间为 2018 年 8 月 7 日，API 等级为 28。

目前移动市场的智能机使用的大部分为 Android 6.0 以上的操作系统，对比以往的版本，该系列版本在系统界面上进行了大幅度的调整，包括应用图标、部件的透明度以及文件夹存储图标的方式，开发者可以下载 Android 6.0 以上的 SDK 来开发和测试。

## 1.1.3　Android 系统架构

Android 的系统架构和其操作系统一样，采用了分层的架构。从架构图看，Android 分为 4 层，从高层到低层分别是应用程序层（Application）、应用程序框架层（Application Framework）、系统运行库层和运行环境层（Libraries＋Android Runtime）和 Linux 内核层（Linux Kernel），具体如图 1-2 所示。

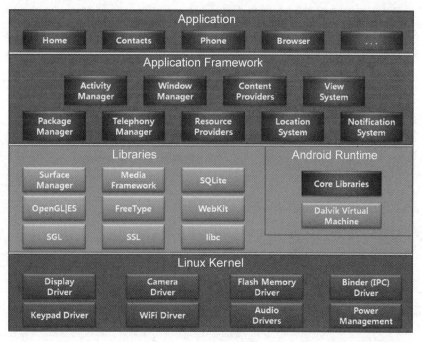

图 1-2　Android 系统架构

接下来将对 Android 的系统架构进行详细的讲解。
- 应用程序层。Android 会同一系列核心应用程序包一起发布，该应用程序包包括客户端、SMS 程序、日历、地图、浏览器、联系人管理程序等。所有的应用程序都是使用 Java 语言编写的。
- 应用程序框架层。应用程序框架提供了大量的 API 供开发者使用。Android 自带的一些核心应用就是使用这些 API 完成的，例如视图（View System）、活动管理器（Activity

Manager)等,开发者也可以通过这些 API 来构建自己的应用程序。除了这些,它也是软件复用的手段,任何应用程序都可以发布它的功能模块,只要遵守了框架约定,那么其他的应用程序就可以使用这个功能模块。
- 系统运行层。其下包括核心类库(Core Libraries)。Android 包含一些 C/C++库,这些库能被 Android 系统中不同的组件使用。它们通过 Android 应用程序框架为开发者提供服务。系统 C 库:一个从 BSD 继承来的标准 C 系统函数库,它是专门为基于嵌入式 Linux 的设备定制的;Surface Manager:对显示子系统的管理,并且为多个应用程序提供了 2D 和 3D 图层的无缝融合等。
- Linux 内核层:Android 是运行于 Linux Kernel 之上,但并不是 GNU/Linux。Android 的 Linux 内核包括显示驱动程序、摄像头驱动程序、Binder(IPC)驱动程序、键盘驱动程序、WiFi 驱动程序、音频驱动程序以及电源管理。

## 1.2 搭建 Android 开发环境

在开始学习 Android 开发之前,学习者应该具备一定的 Java 编程基础,然后再开始学习 Android 的环境搭建和程序开发等。下面将介绍 Android 开发环境的搭建以及调试等。

Android Studio 是 Google 公司为 Android 提供的官方 IDE 工具,Google 公司建议广大 Android 开发者尽快从 Eclipse+ADT 的开发环境改为使用 Android Studio。

Android Studio 不再基于 Eclipse,而是基于 IntelliJ IDEA 的 Android 开发环境。它为 Google 服务和各种设备类型提供扩展模板支持、支持主题编辑的富布局编辑器、可捕捉性能、可用性、版本兼容性以及其他问题的 Lint 工具等。

Android Studio 和 Android SDK 的下载和安装具体步骤如下所示:

(1) 登录 http://www.android-studio.org/页面,网站默认显示 Windows 系统下的版本下载,如图 1-3 所示。

图 1-3 下载 Android Studio

(2) 运行下载后的安装包,3.x 版本的 Android Studio 安装包中组件选择只有两项,去除了 SDK 安装选项,SDK 在 Android Studio 第一次运行时在线下载,如图 1-4 所示。

(3) 第一次运行 Android Studio 时会弹出 Unable to access Android SDK add-on list 提示框,单击 Cancel 按钮,如图 1-5 所示。

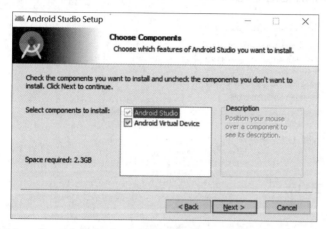

图 1-4　安装 Android Studio

图 1-5　第一次运行 Android Studio 遇到的问题

(4) 之后就是按照向导一路单击 Next 按钮往下走,全部安装完成后出现如图 1-6 所示的界面。

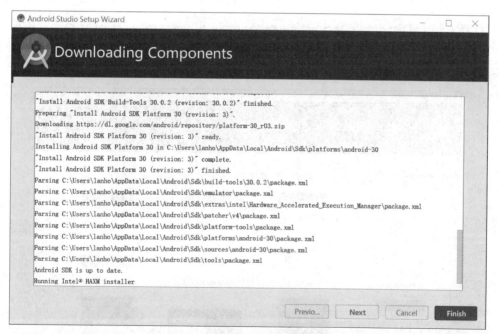

图 1-6　Android Studio 及 SDK 安装完成

(5) 单击 Finish 按钮,开启 Android Studio,接下来可单击 Start a new Android Studio project 来创建一个 Android Studio 项目,具体的界面操作如图 1-7 所示。

图 1-7　新建项目

## 1.3　开发第一个 Android 程序

视频讲解

基本上刚开始学习任何一门语言，第一个程序都是 HelloWorld。本节就教大家开发第一个 Android 程序，并了解 Android 的项目结构。

### 1.3.1　创建与运行第一个 HelloWorld 程序

（1）打开 Android Studio，单击 Start a new Android Studio project，首先选择项目模板，包括应用的平台，例如手机、电视等，然后选择布局界面，通常选择适合自己 App 的界面布局，具体操作如图 1-8 所示。

图 1-8　选择项目模板

（2）设置完成后单击 Next 按钮，出现 Create New Project 对话框，输入项目名称，选择语言和 API 的版本号，如图 1-9 所示。

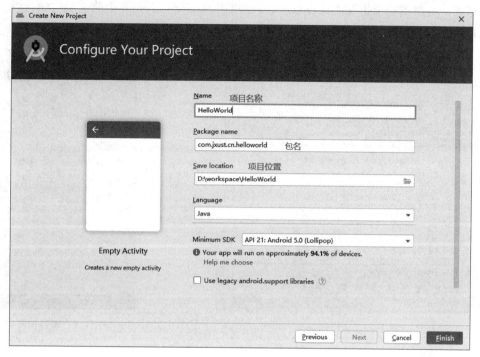

图 1-9　新项目创建

（3）设置完成后单击 Next 按钮，进行 Activity 命名，最后单击 Finish 按钮，出现项目的具体结构图，如图 1-10 所示。

图 1-10　项目结构图

（4）创建 Android 模拟器。单击 Android Studio 中的 AVD Manager 按钮，添加模拟器，具体的操作如图 1-11 所示。

（5）完成上一步创建模拟器的操作以后，可以单击 Android Studio 中的运动按钮，运行项目，具体操作如图 1-12 所示，运行结果如图 1-13 所示。

图 1-11　创建模拟器

图 1-12　运行项目

图 1-13　项目运行图

## 1.3.2　学习项目文件

对于每一个创建成功的 Android 项目，ADT 都会智能地生成两个默认的文件，即布局文件和 Activity 文件。布局文件主要用于展示 Android 项目的界面，Activity 文件主要用于完成界面的交互功能。

Helloworld_activity.xml 的布局文件内容如下所示：

```xml
<RelativeLayout
    xmlns:android = "http://schemas.android.com/apk/res/android"
    xmlns:tools = "http://schemas.android.com/tools"
    android:layout_width = "match_parent"
    android:layout_height = "match_parent"
    tools:context = "com.example.cn.helloworld.MainActivity">
    <TextView
        android:layout_width = "wrap_content"
        android:layout_height = "wrap_content"
        android:text = "Hello World!"/>
</RelativeLayout>
```

在 Helloworld_activity.xml 的布局文件当中可以任意添加 Android 中的组件，还可以更改背景和布局方式等。

HelloWorldActivity 文件内容如下所示：

```
import android.support.v7.app.AppCompatActivity;
import android.os.Bundle;
public class HelloWorldActivity extends Activity {
    @Override
    protected void onCreate(Bundle savedInstanceState) {
        super.onCreate(savedInstanceState);
        setContentView(R.layout.activity_main);
    }
}
```

HelloWorldActivity 继承自 Activity，当执行该类时会先执行 onCreate()方法，然后通过调用 setContentView(R.layout.activity_main)将布局文件转换为 View 对象，通过模拟器显示在界面上。

### 1.3.3　Android 项目结构

在 Android 程序创建完成后，会生成一个基本的项目结构，在开发之前有必要熟练掌握项目结构。接下来具体介绍各个文件。项目结构如图 1-14 所示。

- src：存放项目开发所使用到的 Activity，可以有多个不同的包，在这里 Activity 和普通的 Java 类是一样的。还有各种资源文件（放在 main\\res 子目录下）和 AndroidManifest.xml 文件，除了这些还包含 Android 测试项目。
- res：存放 Android 项目的各种资源文件，例如布局 layout 文件、values 目录下的文件，还存放图片的文件夹 drawable 等。
- libs：存储 Android 项目开发所使用到的第三方 Jar 包。
- build：Android Studio 自动生成的各种源文件，R.java 文件也放在该目录下。

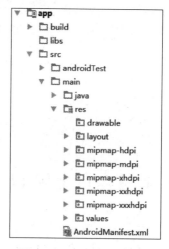

图 1-14　Android 项目结构

### 1.3.4　AndroidManifest.xml 详解

AndroidManifest.xml 清单文件是每个 Android 项目所必需的，它是对整个 Android 应用的全局描述文件。该清单文件详细说明了应用的图标、名称以及包含的各种组件等。该清单文件具体包含的信息如下所示：

- 应用程序的包名，该包名可用于唯一地标识该应用。
- 应用程序所包含的组件，如 Activity、Service、BroadcastReceiver 和 ContentProvider 等。
- 应用程序的版本要求。
- 应用使用到的权限。

AndroidManifest.xml 清单文件具体内容如下所示：

```xml
<?xml version = "1.0" encoding = "utf-8"?>
<manifest xmlns:android = http://schemas.android.com/apk/res/android
    <!-- 应用程序的包名 -->
    package = "com.example.cn.helloworld" >
    <application
        android:allowBackup = "true"
        <!-- 应用程序的图标 -->
        android:icon = "@mipmap/ic_launcher"
        <!-- 应用程序的标签 -->
        android:label = "@string/app_name"
        android:roundIcon = "@mipmap/ic_launcher_round"
        android:supportsRtl = "true"
        android:theme = "@style/AppTheme" >
        <!-- 应用程序的 Activity -->
        <activity android:name = ".MainActivity" >
            <intent - filter >
                <!-- 指定该 Activity 为程序的入口 -->
                <action android:name = "android.intent.action.MAIN" />
                <!-- 指定启动应用时运行该 Activity -->
                <category android:name = "android.intent.category.LAUNCHER" />
            </intent - filter >
        </activity >
    </application >
</manifest >
```

## 1.4 【项目案例1】 "橙一色"App 项目创建

视频讲解

在学习了前面内容之后,大家对安卓开发环境搭建、项目创建等有了初步了解。接下来将介绍一个 App 项目从创建到开发完成所包含的知识点及开发流程。在目前已经具备的知识下,先从创建一个"橙一色"App 项目开始。

### 1. 项目的创建

参考 1.3 节创建 HelloWorld 的流程,创建一个新项目,命名为 OrangeSale_01,后续所有功能的添加都将基于这个项目进行操作,从而最终形成一个完整的项目。

### 2. AndroidManifest.xml 文件的修改

项目创建完成以后,为 App 起个名字叫作"橙一色",该 App 主要实现在线水果的浏览、搜索、购物车添加等功能。因此需要去修改 AndroidManifest.xml 中 App 的名字,具体内容如下:

```xml
<?xml version = "1.0" encoding = "utf-8"?>
<manifest xmlns:android = "http://schemas.android.com/apk/res/android"
```

```
        package = "com.jxust.OrangeSale.cn">
<application
        ...
        android:label = "@string/app_name"
        ...
</application>
</manifest>
```

在 res/values 下的 strings.xml 中修改 app_name，配置内容如下：

```
strings.xml
<resources>
    <string name = "app_name">橙一色</string>
</resources>
```

### 3. 运行效果图

修改完 App 名称以后，再修改一下默认生成的 TextView 的文本内容，改为"脐橙销售"，具体代码不再展示。项目搭建完成以后的运行效果如图 1-15 所示。

图 1-15 "橙一色"App 界面

## 1.5 本章小结

本章主要介绍了 Android 的入门基础知识。首先介绍了 Android 的发展历史，然后介绍了 Android 开发的环境搭建，实现了经典的 HelloWorld 程序，以此讲解 Android 开发的具体步骤，最后引入本书的"橙一色"App 项目案例。本章中开发环境搭建以及如何创建一个安卓项目是开发 Android 所必须要掌握的，要求能熟练掌握。

## 1.6 课后习题

1. 简述各种手机操作系统的特点。
2. 简述 Android 平台的特征。
3. 简述 R.java 和 AndroidManifest.xml 文件的用途。
4. 描述 Android 平台体系结构的层次划分,并说明各个层次的作用。
5. 编写一个 Android 程序并运行。

# Android应用界面

**学习目标**
- 掌握 Android 开发当中常用的 UI 组件的使用。
- 掌握各种布局方式。
- 掌握 ListView 和 Adapter 的使用。

Android 应用开发的一项非常重要的内容就是用户界面开发,一个友好的图形用户界面(Graphics User Interface,GUI)将会对 App 的推广使用起到很关键的作用,会吸引大量的用户使用。

实际上 Android 提供了大量功能全面的 UI 组件,在开发过程当中,只需要把这些组件按照一定的布局方式组合起来,就能构造出一个优秀的功能界面。为了让这些 UI 组件能够响应用户的单击、键盘动作等,Android 也提供了事件响应机制,从而保证了用户与图形界面的交互。

## 2.1 View 概念

Android 应用的绝大部分 UI 组件都放在 android. widget 包及其子包、android. view 包及其子包中,Android 应用的界面都是由 View 和 ViewGroup 对象构建的,它有很多种类,并且都是 View 类的子类。View 类是 Android 系统平台上用户界面的基本单元。ViewGroup 是 View 的一个扩展,可以容纳多个 View,ViewGroup 类可以作为容器来盛装其他组件。Android 图形用户界面的组件层次如图 2-1 所示。

从图 2-1 可以看出,多个视图组件(View)可以存放在一个视图容器(ViewGroup)中,该容器可以与其他视图组件共同存放在另一个视图容器当中,但是一个界面文件必须有且仅有一个容器作为根节点。

View 类是 Android 的一个非常重要的超类,它是 Android 中所有与用户交互的组件的父类,包括 Widget 类的交互 UI 组件(按钮、文本框等)和 ViewGroup 类布

图 2-1 Android 图形用户界面的组件层次

局组件。View 类常用的 XML 属性、相关方法及说明如表 2-1 所示。

表 2-1　View 类常用的 XML 属性、相关方法及说明

| XML 属性 | 方　法 | 说　明 |
| --- | --- | --- |
| android:alpha | setAlpha(float) | 设置透明度 |
| android:background | setBackgroundResource(int) | 设置背景 |
| android:id | setId(int) | 设置组件标识 |
| android:visibility | setVisibility(int) | 设置组件是否可见 |
| android:keepScreenOn | setkeepScreenOn(boolean) | 设置组件是否会强制手机屏幕一直打开 |
| Android:longClickable | setLongClickable(boolean) | 设置是否响应长单击事件 |
| android:scaleX | setScaleX(float) | 设置水平方向的缩放比 |
| android:scaleY | setScaleY(float) | 设置垂直方向的缩放比 |
| android:scrollX | — | 设置水平方向的滚动偏移 |
| android:scrollY | — | 设置垂直方向的滚动偏移 |
| android:scrollbars | — | 定义该组件滚动时显示几个滚动条,该属性支持如下属性值:<br>• none:滚动条不显示。<br>• horizontal:显示水平滚动条。<br>• vertical:显示垂直滚动条 |
| android:rotationX | setRotationX(float) | 设置绕 X 轴旋转的角度 |
| android:rotationY | setRotationY(float) | 设置绕 Y 轴旋转的角度 |
| android:scrollbarStyle | setScrollBarStyle(int) | 设置滚动条风格和位置,该属性具有如下属性值:<br>• insideOverlay。<br>• insideInset。<br>• outsideOverlay。<br>• outsideInset。 |
| android:tag | — | 为该组件设置一个字符串类型的 tag,通过 View 的 getTag()获取该字符串 |
| android:fadeScrollbars | setScrollbarFadingEnabled(boolean) | 当不使用该组件的滚动条时,是否淡出显示滚动条 |
| android:contentDescription | setContentDescription(CharSequence) | 设置该组件的内容描述信息 |
| android:focusable | setFocusable(boolean) | 设置该组件是否可以得到焦点 |
| android:onClick | — | 设置组件的单击事件 |
| android:padding | setPadding(int,int,int,int) | 在组件的四边设置填充区域 |
| android:transformPivotX | setPivotX(float) | 设置该组件旋转时旋转中心的 X 坐标 |
| android:transformPivotY | setPivotY(float) | 设置该组件旋转时旋转中心的 Y 坐标 |
| android:translationX | setTranslationX(float) | 设置该组件在 X 方向上的位移 |
| android:translationY | setTranslationY(float) | 设置该组件在 Y 方向上的位移 |
| android:translationZ | setTranslationZ(float) | 设置该组件在 Z(垂直屏幕)方向上的位移 |
| android:soundEffectsEnabled | setSoundEffectsEnabled(boolean) | 设置该组件被单击时,是否使用音效 |
| android:minHeight | setMinimumHeight(int) | 该组件的最小高度 |
| android:minWidth | setMinimumWidth(int) | 该组件的最小宽度 |

## 2.2 布局管理器

在 Android 开发中,界面的设计是通过布局文件实现的,布局文件采用 XML 的格式,每个应用程序默认都会创建一个 activity_main.xml 布局文件,它是应用启动的界面。接下来将详细介绍布局的创建使用以及布局的类型。

### 2.2.1 创建和使用布局文件

在开始一个新的界面设计时,需要重新创建一个新的布局文件,下面将详细介绍如何创建一个新的布局文件以及如何在 Activity 中使用。

(1) 打开项目,找到 layout 文件夹,如图 2-2 所示,右击,选择 New→XML→Layout XML File(布局文件)命令,然后就会创建一个新的布局文件。

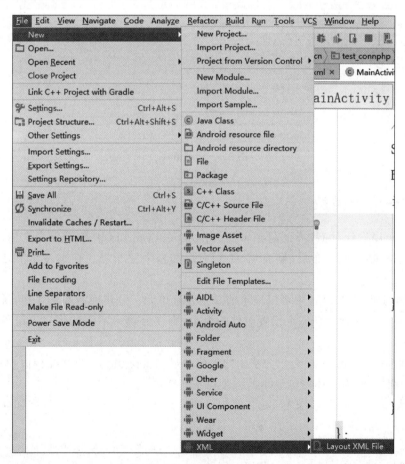

图 2-2 创建布局文件

(2) 新创建的布局文件可以通过在 XML 文件中添加组件,也可以通过在图形用户界面上进行直接的拖动操作,然后再次通过代码进行调整,这种方式减少了用户的代码编写量。具体的操作如

图 2-3 所示。

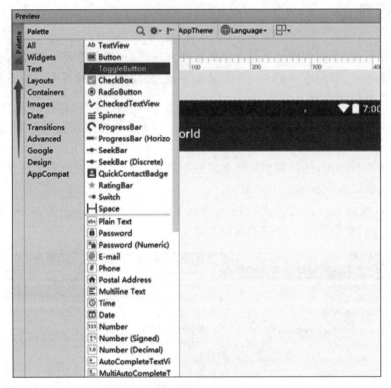

图 2-3　拖动组件

## 2.2.2　布局的类型

一个优秀的布局设计对 UI 界面起到重要的作用,在 Android 中布局分为 7 种,分别是相对布局、线性布局、表格布局、网格布局、帧布局、绝对布局和扁平化布局。在学习本节之前先介绍基本的宽、高单位。

- px：代表像素,即在屏幕中可以显示的最小元素单元。分辨率越高的手机,屏幕的像素点就越多。因此,如果使用 px 来设置组件的大小,则在分辨率不同的手机上组件显示出来的大小也不一样。
- pt：代表磅数,一般 pt 都会作为字体的单位来显示。pt 和 px 的情况相似,在不同分辨率的手机上,用 pt 作为字体单位显示的大小也不一样。
- dp：代表密度无关像素,又称 dip。使用 dp 的好处是无论屏幕的分辨率如何总能显示相同的大小,一般使用 dp 作为组件与布局的宽高单位。
- sp：代表可伸缩像素,采用与 dp 相同的设计理念,设置字体大小时使用。

下面将通过示例详细介绍 7 种布局的用法。

### 1. 相对布局

相对布局(RelativeLayout)分为相对于容器和组件两种。为了更好地确定布局中组件的位置,相对布局提供了很多属性，具体如表 2-2 所示。

表 2-2　相对布局中组件属性描述

| 属 性 声 明 | 功 能 描 述 |
| --- | --- |
| android:layout_alignParentLeft | 是否跟父布局左对齐 |
| android:layout_alignParentRight | 是否跟父布局右对齐 |
| android:layout_alignParentTop | 是否跟父布局顶部对齐 |
| android:layout_alignParentBottom | 是否跟父布局底部对齐 |
| android:layout_toRightOf | 在指定组件右边 |
| android:layout_toLeftOf | 在指定组件左边 |
| android:layout_above | 在指定组件上边 |
| android:layout_below | 在指定组件下边 |
| android:layout_alignBaseline | 与指定组件水平对齐 |
| android:layout_alignLeft | 与指定组件左对齐 |
| android:layout_alignRight | 与指定组件右对齐 |
| android:layout_alignTop | 与指定组件顶部对齐 |
| android:layout_alignBottom | 与指定组件底部对齐 |

表 2-2 介绍了相对布局中组件的属性描述，下面将通过一个相对布局界面来具体介绍如何使用相对布局。

图 2-4 是用相对布局的属性来控制组件的大小和位置，具体代码如下所示：

```xml
<?xml version = "1.0" encoding = "utf-8"?>
< RelativeLayout xmlns:android = "http://schemas.android.com/apk/res/android"
    xmlns:tools = "http://schemas.android.com/tools"
    android:layout_width = "match_parent"
    android:layout_height = "match_parent"
    tools:context = "com.jxust.cn.relativelayout.MainActivity">
    < Button
        android:id = "@ + id/button"
        android:layout_width = "wrap_content"
        android:layout_height = "wrap_content"
        android:layout_alignParentTop = "true"
        android:layout_marginTop = "100dp"
        android:layout_marginLeft = "50dp"
        android:text = "Button1" />
    < Button
        android:id = "@ + id/button2"
        android:layout_width = "wrap_content"
        android:layout_height = "wrap_content"
        android:layout_toRightOf = "@ + id/button"
        android:layout_below = "@ + id/button"
        android:layout_marginTop = "15dp"
        android:text = "Button2"/>
</RelativeLayout>
```

上述代码中,通过使用<RelativeLayout>标签来定义一个相对布局,并且添加了两个Button组件。其中BUTTON1通过alignParentTop="true"来指定位于屏幕顶部,通过layout_marginTop="100dp"来设置距离顶部的距离。BUTTON2通过layout_toRightOf="@+id/button"来设置位于BUTTON1右边,通过layout_below来设置位于BUTTON1下边,然后又通过layout_margintop设置距离BUTTON1的垂直距离,如图2-4所示。

### 2. 线性布局

线性布局(LinearLayout)是Android中较为常用的布局方式,它使用<LinearLayout>标签,主要分为垂直线性布局和水平线性布局,如图2-5所示。

图2-4 相对布局

(a) 垂直线性布局

(b) 水平线性布局

图2-5 线性布局

上述线性布局界面对应的部分代码如下所示:

```xml
<LinearLayout xmlns:android = "http://schemas.android.com/apk/res/android"
    android:layout_width = "match_parent"
    android:layout_height = "match_parent"
    android:orientation = "horizontal">
    <Button
        android:id = "@ + id/button1"
        android:layout_width = "wrap_content"
        android:layout_height = "wrap_content"
        android:text = "Button1" />
    <Button
        android:id = "@ + id/button2"
        android:layout_width = "wrap_content"
        android:layout_height = "wrap_content"
        android:text = "Button2" />
    <Button ..../>
</LinearLayout>
```

上述代码使用线性布局放了3个按钮,布局是水平还是垂直取决于android:orientation,该属性的取值有vertical(垂直)和horizontal(水平)两种。

### 3. 表格布局

表格布局(TableLayout)的方式不同于前面两种，它是让组件以表格的形式来排列组件，只要将组件放在单元格中，组件就可以整齐地排列。

表格布局使用< TableLayout >标签，行数由 TableRow 对象控制，每行可以放多个组件，列数由最宽的单元格决定，假设第一行有 2 个组件，第二行有 3 个组件，那么这个表格布局就有 3 列。在组件中使用 layout_column 属性指定具体的列数，该属性的值从 0 开始，代表第一列。下面将通过一个具体的实例来讲解 TableLayout 的使用，如图 2-6 所示。

图 2-6 表格布局

图 2-6 的界面具体实现代码如下：

```xml
< TableLayout xmlns:android = "http://schemas.android.com/apk/res/android"
    android:layout_width = "match_parent"
    android:layout_height = "match_parent"
    android:stretchColumns = "1">
    < TableRow
        android:layout_width = "match_parent"
        android:layout_height = "match_parent">
        < Button
            android:id = "@ + id/button1"
            android:layout_width = "wrap_content"
            android:layout_height = "wrap_content"
            android:layout_column = "0"
            android:text = "Button1" />
    </TableRow>
    < TableRow
        android:layout_width = "match_parent"
        android:layout_height = "match_parent">
        < Button
            android:id = "@ + id/button2"
            android:layout_width = "wrap_content"
            android:layout_height = "wrap_content"
            android:layout_column = "1"/>
    </TableRow>
</TableLayout >
```

上述代码中使用表格布局的方式，设计了两行，每行一个按钮的布局，使用 android:stretchColumns="1"属性拉伸第二列，android:layout_column="0"属性表示显示在第一列中。

需要注意的是，TableRow 不需要设置宽度和高度，其宽度一定是自动填充容器，高度根据内容改变。但对于 TableRow 的其他组件来说，是可以设置宽度和高度的，但必须是 wrap_content 或者 fill_parent。

#### 4. 网格布局

网格布局(GridLayout)是 Android 4.0 新增的布局管理器,因此需要在 4.0 之后的版本才能使用。它的作用类似于 table,它把整个容器划分为 rows×columns 个网格,每个网格可以放置一个组件。GridLayout 提供了 setRowCount(int)和 setColumnCount(int)方法来控制该网格的行数量和列数量。下面通过具体的实例讲解网格布局的使用,如图 2-7 所示。

图 2-7 网格布局

图 2-7 的界面具体实现代码如下:

```xml
<?xml version = "1.0" encoding = "utf-8"?>
<GridLayout xmlns:android = "http://schemas.android.com/apk/res/android"
    android:layout_width = "wrap_content"
    android:layout_height = "wrap_content"
    android:layout_gravity = "center"
    android:columnCount = "4"
    android:rowCount = "4"
    android:orientation = "horizontal">
    <Button android:text = "/"
        android:layout_column = "3"/>
    <Button android:text = "1"/>
    <Button android:text = "2"/>
    <Button android:text = "3"/>
    <Button android:text = " * "/>
    <Button android:text = "4"/>
    <Button android:text = "5"/>
    <Button android:text = "6"/>
    <Button android:text = " - "/>
    <Button android:text = "7"/>
    <Button android:text = "8"/>
    <Button android:text = "9"/>
    <Button android:text = " + "
        android:layout_gravity = "fill"
```

```
        android:layout_rowSpan = "2"/>
    < Button android:text = "0"/>
    < Button android:text = " = "
        android:layout_gravity = "fill"
        android:layout_columnSpan = "2" />
</GridLayout>
```

上述代码采用了网格布局 GridLayout 设计了一个简单的计算器界面，分别使用 columnCount 和 rowCount 属性设置整体界面布局为 4 行 4 列，其中"＝"按钮通过设置属性 columnSpan＝"2"表示占据了两列，"＋"按钮通过设置属性使用 rowSpan＝"2"占据了两行。使用 layout_column 表示该按钮在第几列。

### 5. 帧布局

帧布局（FrameLayout）是 Android 布局中最简单的一种，帧布局为每个加入其中的组件都创建了一块空白区域。采用帧布局的方式设计界面时，只能在屏幕左上角显示一个组件，如果添加多个组件，则这些会依次重叠在屏幕左上角显示，且会透明显示之前的文本，如图 2-8 所示。

从图 2-8 可以看到，界面中放了 3 个 Button 组件，最先添加的 BUTTON1 最大，在最下边；然后添加的 BUTTON2 较小一点，在 BUTTON1 上面；最后添加的最小的 BUTTON3 在最上面，这 3 个组件重叠显示在屏幕的左上角。

图 2-8　帧布局

图 2-8 所示帧布局的界面实现代码如下：

```
< FrameLayout xmlns:android = "http://schemas.android.com/apk/res/android"
    android:layout_width = "match_parent"
    android:layout_height = "match_parent">
    < Button
        android:id = "@ + id/button10"
        android:layout_width = "314dp"
        android:layout_height = "315dp"
        android:text = "Button1" />
    < Button
        android:id = "@ + id/button11"
        android:layout_width = "216dp"
        android:layout_height = "140dp"
        android:text = "Button2" />
    < Button
        android:id = "@ + id/button12"
        android:layout_width = "103dp"
        android:layout_height = "wrap_content"
        android:text = "Button3" />
</FrameLayout>
```

### 6. 绝对布局

绝对布局(AbsoluteLayout)是通过指定 x、y 坐标来控制每一个组件的位置,放入该布局的组件需要通过 android:layout_x 和 android:layout_y 两个属性指定其在屏幕上的确切位置。把屏幕看作一个坐标轴,左上角为(0,0),往屏幕下方为 y 正半轴,右方为 x 正半轴。下面将通过一个具体的例子来讲解绝对布局的使用过程,如图 2-9 所示。

图 2-9 绝对布局

从图 2-9 可以看出,布局里放置了一个 Button 组件,然后通过指定其 x 和 y 坐标来放置其位置。具体实现代码如下:

```xml
<AbsoluteLayout xmlns:android="http://schemas.android.com/apk/res/android"
    android:layout_width="match_parent"
    android:layout_height="match_parent">
    <Button
        android:id="@+id/button3"
        android:layout_width="wrap_content"
        android:layout_height="wrap_content"
        android:layout_x="71dp"
        android:layout_y="25dp"
        android:text="Button" />
</AbsoluteLayout>
```

理论上绝对布局可以设计任何布局,但是实际的工程当中不提倡使用这种布局,因为在使用这种布局时需要精确地计算各个组件的大小,而且运行在不同大小的屏幕上产生的效果也不相同。

### 7. 扁平化布局

ConstraintLayout 是 Android Studio 2.2 中主要的新增功能之一,也是 Google 在 2016 年的 I/O 大会上重点宣传的一个功能。目前 Android Studio 创建新项目生成的默认布局就是扁平化布局(ConstraintLayout)。在传统的 Android 开发中,界面基本都是靠编写 XML 代码完成的,虽然 Android Studio 也支持可视化的方式来编写界面,但是操作起来并不方便,而 ConstraintLayout 就是为了解决这一现状而出现的。它和传统编写界面的方式恰恰相反,ConstraintLayout 非常适合使用可视化的方式来编写界面,但并不太适合使用 XML 的方式来进行编写。当然,可视化操作的背后仍然还是使用 XML 代码来实现,只不过这些代码是由 Android Studio 根据我们的操作自动生成的。

Android Studio 默认生成的 ConstraintLayout 布局代码如下所示：

```xml
<?xml version = "1.0" encoding = "utf-8"?>
    < android.support.constraint.ConstraintLayout
xmlns:android = "http://schemas.android.com/apk/res/android"
    xmlns:app = "http://schemas.android.com/apk/res-auto"
    android:id = "@ + id/lay_root"
    android:layout_width = "match_parent"
    android:layout_height = "match_parent">
    < TextView
        android:id = "@ + id/text1"
        android:layout_width = "wrap_content"
        android:layout_height = "wrap_content"
        android:text = "test" />
</android.support.constraint.ConstraintLayout >
```

## 2.3 Android 控件详解

视频讲解

学习完 Android 的布局方式以后，要进行 UI 界面的设计，还需要熟练掌握各种控件的应用。开始一个界面的设计都是先创建容器，然后不断地向容器中添加组件，最后形成一个 UI 界面。掌握这些基本用户界面组件是学好 Android 编程的基础。接下来将详细地介绍各个组件的使用方法。

### 2.3.1 TextView

TextView(文本框)直接继承了 View，它还是 EditText 和 Button 两个 UI 组件类的父类。TextView 的作用就是在界面上显示文字，通过在布局文件中或者在 Activity 中修改文字的内容。下面将通过具体的例子讲解 TextView 的使用，如图 2-10 所示。

图 2-10  TextView 组件

图 2-10 所示界面的实现代码如下：

```xml
< TextView
    android:id = "@ + id/textView"
    android:layout_width = "wrap_content"
    android:layout_height = "wrap_content"
    android:layout_weight = "1"
    android:textColor = "@android:color/black"
    android:textSize = "16sp"
    android:gravity = "center"
    android:layout_marginTop = "120dp"
    android:text = "我爱 Android" />
```

上述代码中放置了一个 TextView 组件，设置文本的宽、高都为适配内容，通过 textColor 设置字体颜色、textSize 设置字体的大小、gravity 设置字体居中显示，更多的属性可以查看 Android 开发文档。

## 2.3.2　EditText

EditText(输入框)与TextView非常相似,许多XML属性都能共用,与TextView的最大区别就是EditText能够接受用户的输入。EditText的重要属性就是inputType,该属性相当于Html的<input.../>元素的type属性,用于将EditText设置为指定类型的输入组件,如手机号、密码、日期等。还有一个属性是当提示用户当前文本框要输入的内容是什么时,使用android:hint=""来提示用户,当用户单击文本框时这些文字就会消失。下面将通过具体的实例来介绍EditText的使用过程,如图2-11所示。

图2-11所示界面的实现代码如下:

```
<EditText
    android:id="@+id/editText"
    android:layout_width="200dp"
    android:layout_height="wrap_content"
    android:layout_gravity="center"
    android:hint="请输入密码"
    android:textColor="@android:color/darker_gray"
    android:textSize="16sp"
    android:inputType="numberPassword" />
```

上述代码通过android:inputType属性来设置输入的类型,通过android:hint属性提示用户当前文本框该输入什么内容。

EditText还可以通过自定义样式的方式来修改组件样式,自定义样式修改结果如图2-12所示。

图2-11　EditText组件

图2-12　自定义EditText样式

图2-12所示的自定义样式实现代码如下:

```
<?xml version="1.0" encoding="UTF-8"?>
<shape xmlns:android="http://schemas.android.com/apk/res/android">
    <solid android:color="#FFFFF" />
```

```
< corners android:radius = "3dip"/>
< stroke
    android:width = "2dip"
    android:color = "♯3F51B5" />
</shape >
<!-- 通过在 EditText 中 background 属性调用 -->
```

### 2.3.3 Button

Button(按钮)继承了 TextView,它主要是 UI 界面上生成的一个按钮,用户可以单击按钮,并且能为按钮添加 onClick 事件,即单击事件。按钮使用起来相对容易,可以通过 android:background 为按钮设置背景或者自定义样式,Button 的 XML 属性和 TextView 相似,大多数属性能够共用。下面将通过具体的实例讲解 Button 的使用。普通 Button 以及自定义 Button 如图 2-13 所示。

(a) 普通Button　　　　　(b) 自定义Button

图 2-13　Button 组件

图 2-13 界面样式的具体实现代码如下:

```
<?xml version = "1.0" encoding = "utf-8"?>
< LinearLayout xmlns:android = "http://schemas.android.com/apk/res/android"
    android:layout_width = "match_parent"
    android:layout_height = "match_parent">
< Button
    android:id = "@ + id/button3"
    android:layout_width = "wrap_content"
    android:layout_height = "60dp"
    android:layout_weight = "1"
    android:textSize = "18sp"
    <!-- 通过在 EditText 中 background 属性调用显示自定义布局样式 -->
    android:background = "@drawable/button_shape"
    android:textColor = "@android:color/white"
    android:text = "Button" />
</LinearLayout>
```

上述是Button组件在布局文件的定义,需要熟练掌握,关于Button组件的单击事件在第3章中会做详细的介绍。

### 2.3.4 ImageView

ImageView(图像视图)继承自View组件,它的主要功能是用于显示图片,除此之外,ImageView还派生了ImageButton、ZoomButton等组件,因此ImageView支持的XML属性和方法基本上也可以应用于ImageButton、ZoomButton等组件。下面将通过具体的例子显示ImageView最基本的使用,如图2-14所示。

(a) ImageView　　　　(b) ImageButton

图 2-14　ImageView 和 ImageButton

图 2-14 所示界面的具体实现代码如下所示:

```xml
<LinearLayout xmlns:android="http://schemas.android.com/apk/res/android"
    xmlns:app="http://schemas.android.com/apk/res-auto"
    android:layout_width="match_parent"
    android:layout_height="match_parent"
    android:orientation="vertical">
    <ImageView
        android:id="@+id/imageView"
        android:layout_width="wrap_content"
        android:layout_height="wrap_content"
        android:layout_gravity="center"
        android:src="@mipmap/ic_launcher" />
    <ImageButton
        android:id="@+id/imageButton"
        android:layout_width="fill_parent"
        android:layout_height="wrap_content"
        android:src="@mipmap/ic_launcher" />
</LinearLayout>
```

上述代码介绍了ImageView和ImageButton组件在XML文件里的基本定义使用,通过android:src设置ImageView的图片来源和ImageButton的背景。除此之外,ImageView和ImageButton也能添加单击事件,这些在以后章节的学习中会详细介绍。

## 2.3.5　RadioButton 和 CheckBox

RadioButton(单选按钮)和 CheckBox(复选框)是用户界面中最普通的 UI 组件,它们都继承自 Button 类,因此可以直接使用 Button 支持的各种属性和方法。

RadioButton 和 CheckBox 都多了一个可选中的功能,因此可以额外指定一个 android:checked 属性,用于指定 RadioButton 和 CheckButton 初始时是否被选中。

RadioButton 和 CheckBox 的不同之处在于,一组 RadioButton 只能选中其中一个,因此 RadioButton 通常要和 RadioGroup 一起使用,用于定义一组单选按钮。

下面将通过具体的实例来学习 RadioButton 和 CheckBox 的使用,实现效果如图 2-15 所示。

(a) RadioButton　　　　　(b) CheckBox

图 2-15　RadioButton 和 CheckBox

图 2-15 的单选按钮和复选框实现代码如下:

```
< LinearLayout xmlns:android = "http://schemas.android.com/apk/res/android"
    android:layout_width = "match_parent"
    android:layout_height = "match_parent"
    android:orientation = "vertical">
    <!-- 单选按钮 -->
    < RadioGroup
    android:layout_width = "match_parent"
    android:layout_height = "wrap_content">
    < RadioButton
        android:id = "@ + id/radioButton2"
        android:layout_width = "wrap_content"
        android:layout_height = "wrap_content"
        android:checked = "true"
        android:text = "男" />
    < RadioButton
        android:id = "@ + id/radioButton"
        android:layout_width = "wrap_content"
        android:layout_height = "wrap_content"
```

```xml
            android:text = "女" />
    </RadioGroup>
    <!-- 复选框 -->
    <CheckBox
        android:id = "@ + id/checkBox"
        android:layout_width = "match_parent"
        android:layout_height = "wrap_content"
        android:text = "读书" />
    <CheckBox
        android:id = "@ + id/checkBox2"
        android:layout_width = "match_parent"
        android:layout_height = "wrap_content"
        android:text = "看电影" />
    <CheckBox
        android:id = "@ + id/checkBox3"
        android:layout_width = "match_parent"
        android:layout_height = "wrap_content"
        android:text = "打篮球" />
    <CheckBox
        android:id = "@ + id/checkBox4"
        android:layout_width = "match_parent"
        android:layout_height = "wrap_content"
        android:checked = "true"
        android:text = "听音乐" />
</LinearLayout>
```

上述代码定义了一个 RadioGroup,其中包含了两个 RadioButton,然后使用 android:checked 属性来设置"男"默认选中,接下来又定义了 4 个复选框,同样使用 android:checked 来选中"读书"和"听音乐"。对这些组件同样可以添加事件处理,例如选中以后改变事件处理等,具体会在以后的章节介绍。

### 2.3.6 ProgressBar

ProgressBar(进度条)也是一组重要的组件,ProgressBar 本身代表了进度条组件,它还派生了两个常用的组件:SeekBar 和 RatingBar。ProgressBar 及其子类十分相似,只是在显示上有一定的区别。ProgressBar 支持的常用属性如表 2-3 所示。

表 2-3 ProgressBar 支持的常用属性

| XML 属性 | 说 明 |
| --- | --- |
| android:max | 设置该进度条的最大值 |
| android:progress | 设置该进度条的已完成进度值 |
| android:progressDrawable | 设置该进度条的轨道对应的 Drawable 对象 |
| android:indeterminate | 设置进度条是否精确显示进度 |
| android:indeterminateDrawable | 设置不显示进度条的 Drawable 对象 |
| android:indeterminateDuration | 设置不精确显示进度的持续事件 |

进度条通常用于向用户显示某个耗时操作完成的百分比。进度条可以动态地显示进度，因此避免了长时间地执行某个耗时操作时，让用户感觉程序失去了响应，从而带给用户更好的体验。

Android 支持多种风格的进度条，通过 style 属性可以为 ProgressBar 指定风格，该属性支持的属性值如表 2-4 所示。

表 2-4　style 属性支持的属性值

| 属 性 值 | 说　　明 |
| --- | --- |
| Widget.ProgressBar.Horizontal | 水平进度条 |
| Widget.ProgressBar.Inverse | 普通大小的环形进度条 |
| Widget.ProgressBar.Large | 大环形进度条 |
| Widget.ProgressBar.Large.Inverse | 不断跳跃、旋转动画的大进度条 |
| Widget.ProgressBar.Small | 小环形进度条 |
| Widget.ProgressBar.Small.Inverse | 不断跳跃、旋转动画的小进度条 |

所以在设计 UI 用户界面的时候，用户可以根据需要选择水平进度条或者环形进度条，也可以根据需要自定义适合项目需要的进度条。ProgressBar 提供了两种方法来操作进度：一种是 setProgress(int)，用来设置进度的完成百分比；另一种是 incrementProgressBy(int)，用来设置进度条增加或减少。当参数为正时，进度增加，反之相反。下面将通过具体的实例学习进度条的使用，如图 2-16 所示。

(a) 环形进度条　　　　　(b) 水平进度条

图 2-16　环形进度条与水平进度条

图 2-16 所示界面的布局实现代码如下：

```
<?xml version = "1.0" encoding = "utf-8"?>
< LinearLayout xmlns:android = "http://schemas.android.com/apk/res/android"
    android:layout_width = "match_parent"
    android:layout_height = "match_parent"
```

```
        android:orientation = "vertical">
    < ProgressBar
        android:id = "@ + id/progressBar5"
        style = "?android:attr/progressBarStyle"
        android:layout_width = "match_parent"
        android:layout_height = "wrap_content" />
    < ProgressBar
        android:id = "@ + id/progressBar4"
        style = "?android:attr/progressBarStyleHorizontal"
        android:layout_width = "145dp"
        android:layout_height = "25dp"
        android:layout_gravity = "center"
        android:layout_marginTop = "30dp"
        android:background = "@android:color/holo_green_light" />
</LinearLayout >
```

从上面的代码中,可以看到使用了线性垂直布局,界面当中放了两个进度条,分别为环形进度条和水平进度条,代码中使用了 style 属性来设置进度条的样式,除此之外,还可以通过 android:background 设置进度条的背景颜色或者自定义布局文件,同前面讲述的 Button 和 EditText 自定义样式使用方式相同,只需要修改颜色等属性搭配布局即可。关于 ProgressBar 的使用方式,后续会结合具体的案例进行讲解。

### 2.3.7 SeekBar

SeekBar(拖动条)与进度条非常相似,只是进度条采用颜色填充来表示进度完成的程度,而拖动条则通过滑块的位置来标识数字。拖动条允许用户拖动滑块来改变值,因此拖动条通常用于对系统的某种数值进行调节,比如音量调节等。

图 2-17  SeekBar 组件

由于拖动条继承了进度条,因此进度条所支持的 XML 属性和方法同样适用于拖动条。进度条允许用户改变拖动条的滑块外观,改变滑块外观通过 android:thumb 属性来指定,这个属性指定一个 Drawable 对象,该对象将作为自定义滑块。为了让程序能够响应拖动条滑块位置的改变,程序可以为它绑定一个 OnSeekBarChangeListener 监听器。下面将通过具体的实例讲解 SeekBar 的使用,如图 2-17 所示。

图 2-17 所示 SeekBar 的界面布局实现代码如下:

```
<?xml version = "1.0" encoding = "utf-8"?>
< LinearLayout xmlns:android = "http://schemas.android.com/apk/res/android"
    android:layout_width = "match_parent"
    android:layout_height = "match_parent">
    < SeekBar
```

```
            android:id = "@ + id/seekBar"
            android:layout_width = "wrap_content"
            android:layout_height = "wrap_content"
            android:layout_marginTop = "20dp"
            android:background = "@android:color/holo_green_light"
            android:layout_weight = "1" />
</LinearLayout>
```

从上述代码中可以看出,使用了线性布局,并在当中加入了一个拖动条,通过 android:background 属性设置了拖动条的背景样式。关于拖动条的拖动事件在后续章节中将会通过具体项目案例进行讲解。

上面几小节介绍了 TextView、EditText、Button、ImageView、RadioButton、CheckBox、ProgressBar 以及 SeekBar 的基本用法,讲述了如何在布局文件中定义以及如何自定义样式等。关于组件的单击事件等会在后续章节中详细讲解。

## 2.4 AdapterView 及其子类

AdapterView 是一种重要的组件。AdapterView 本身是一个抽象基类,它派生的子类在用法上与其十分相似,只是显示界面上有一定的区别。AdapterView 具有如下特征:
- AdapterView 继承了 ViewGroup,它的本质是容器。
- AdapterView 可以包括多个"列表项",并以合适的方式显示出来。
- AdapterView 显示的多个"列表项"由 Adapter 提供。调用 AdapterView 的 setAdapter (Adapter)方法设置 Adapter 即可。

### 2.4.1 ListView 和 ListActivity

视频讲解

ListView 是手机系统中使用非常广泛的一种组件,它以垂直列表的形式显示所有的列表项。生成列表视图有如下两种方式:
- 直接使用 ListView 进行创建。
- 创建一个继承 ListActivity 的 Activity(相当于该 Activity 显示的组件为 ListView)。

一旦在程序中获得了 ListView 之后,接下来就需要为 ListView 设置它要显示的列表项了。在这一点上,ListView 显示出了 AdapterView 的特征:通过 setAdapter(Adapter)方法为之提供 Adapter,并由 Adapter 提供列表项即可。

ListView 提供的常用 XML 属性如表 2-5 所示。

表 2-5 ListView 的常用 XML 属性

| XML 属性 | 说 明 |
| --- | --- |
| android:divider | 设置分割条样式(颜色或者 Drawable 对象) |
| android:dividerHeight | 设置分割条高度 |
| android:entries | 指定一个数组资源,用来填充 ListView 项 |

续表

| XML 属性 | 说 明 |
| --- | --- |
| android:footerDividersEnabled | 设置为 false,则不在 FooterView 之前绘制分割条 |
| android:headerDividersEnabled | 设置为 false,则不在 HeaderView 之后绘制分割条 |
| android:scrollbars | 设置是否显示滚动条 |
| android:fadingEdge | 设置是否去除 ListView 滑到顶部和底部时边缘的黑色阴影 |
| android:listSelector | 设置是否去除单击颜色 |
| android:cacheColorHint | 设置 ListView 去除滑动颜色 |

图 2-18　ListView 组件

下面将通过一个基于数组的 ListView 实例来讲解 ListView 的使用,先看运行图,如图 2-18 所示。

要实现上面的界面,首先需要在布局文件中添加 ListView 组件,然后在 values 文件夹下添加一个新的 arrays.xml 文件用来存储数组元素,在 ListView 组件中通过 android:entries 调用,把数组元素加载到 listview 中。具体的实现代码如下:

(1) 布局文件代码。

```xml
<?xml version = "1.0" encoding = "utf-8"?>
<LinearLayout xmlns:android = "http://schemas.android.com/apk/res/android"
    android:layout_width = "match_parent"
    android:layout_height = "match_parent">
<!-- 直接使用数组资源给 ListView 添加列表项 -->
    <!-- 设置分割条的颜色 -->
    <!-- 设置分割条的高度 -->
    <ListView
        android:id = "@ + id/listview1"
        android:layout_width = "match_parent"
        android:layout_height = "wrap_content"
        android:divider = "#C4C4C4"
        android:entries = "@array/teacher_name"
        android:dividerHeight = "1dp">
    </ListView>
</LinearLayout>
```

(2) values 下的 arrays.xml 文件代码。

```xml
<?xml version = "1.0" encoding = "utf-8"?>
<resources>
    <!-- 添加数组元素 -->
    <string-array name = "teacher_name">
        <item>张三</item>
        <item>李四</item>
        <item>王五</item>
        <item>赵六</item>
```

```
</string-array>
</resources>
```

使用数组创建 ListView 是一种非常简单的方式,但是这种 ListView 能定制的内容很少,如果想对 ListView 的外观、行为等进行自定义,就需要把 ListView 作为 AdapterView 使用,通过 Adapter 自定义每个列表项的外观、内容以及添加的动作行为等。

### 2.4.2 Adapter 接口

视频讲解

Adapter 本身只是一个接口,它派生了 ListAdapter 和 SpinnerAdapter 两个子接口,其中 ListAdapter 为 AbsListView 提供列表项,而 SpinnerAdapter 为 AbsSpinner 提供列表项。

Adapter 常用的实现类如下所示:
- BaseAdapter——一个抽象类,继承它需要实现较多的方法,所以也就具有较高的灵活性。
- ArrayAdapter——支持泛型操作,最为简单,只能展示一行字。
- SimpleAdapter——有最好的扩充性,可以自定义各种效果。

下面将通过具体的实例讲解常用的 Adapter 实现类的用法。

**实例一:基于 ArrayAdapter 创建 ListView**
- 布局界面代码同基于数组创建 ListView 的布局代码一样。
- Activity 代码如下:

```
public class MainActivity extends AppCompatActivity {
    @Override
    protected void onCreate(Bundle savedInstanceState) {
        super.onCreate(savedInstanceState);
        setContentView(R.layout.listview_layout);
        ListView listView = (ListView)findViewById(R.id.listview1);
        //定义一个数组,用来填充 ListView
        String[] arr = {"章节 1","章节 2","章节 3"};
        ArrayAdapter<String> adapter = newArrayAdapter<String>
                (this,android.R.layout.simple_expandable_list_item_1,arr);
        //为 ListView 设置 Adapter
        listView.setAdapter(adapter);
    }
}
```

实现效果如图 2-19 所示。

从上面加粗代码中可以看到创建了一个 ArrayAdapter,创建 ArrayAdapter 时必须要指定 3 个参数,分别为 Context、textViewResourceId、数组或 List。第一个参数代表了访问整个 Android 应用的接口;第二个参数表示一个资源 ID,该资源 ID 代表一个 TextView,负责设置列表项的样式,可以自定义;第三个参数代表加入列表项的元素。

图 2-19 基于 ArrayAdapter 实现

**实例二：基于 SimpleAdapter 创建 ListView**

使用 ArrayAdapter 实现 Adapter 虽然比较简单,但是只能实现比较单一的列表,即每个列表项只能是 TextView,如果开发者考虑在每一行放置不同的组件,则可以考虑使用 SimpleAdapter。下面将通过一个实例讲解 SimpleAdapter 的使用。

- 主布局界面代码同基于数组创建 ListView 的布局代码一样。
- 使用 SimpleAdapter,需要在 layout 目录下添加一个自定义的布局文件 list_item_layout.xml,即每一行的布局样式,代码如下：

```xml
<LinearLayout xmlns:android = "http://schemas.android.com/apk/res/android"
    android:layout_width = "match_parent"
    android:layout_height = "match_parent"
    android:orientation = "horizontal">
    <!-- 定义一个 ImageView 组件,用来显示头像 -->
    <ImageView
        android:id = "@ + id/icon"
        android:layout_width = "wrap_content"
        android:layout_height = "wrap_content" />
    <LinearLayout
        android:layout_width = "match_parent"
        android:layout_height = "wrap_content"
        android:orientation = "vertical">
        <!-- 定义一个 TextView 组件,用来显示名字 -->
        <TextView
            android:id = "@ + id/name"
            android:layout_width = "wrap_content"
            android:layout_height = "wrap_content"
            android:textSize = "16sp"/>
        <!-- 定义一个 TextView 组件,用来显示人物的描述 -->
        <TextView
            android:id = "@ + id/dexc"
            android:layout_width = "wrap_content"
            android:layout_height = "wrap_content"
            android:textSize = "16sp"/>
    </LinearLayout>
</LinearLayout>
```

- Activity 代码如下：

```java
public class MainActivity extends Activity {
    //定义名字数组
    private String[] name = {"张三", "王五", "赵六"};
```

```
//定义描述任务数组
private String[] desc = {"唱歌","跳舞","打球"};
//定义头像数组
private int[] icon = new int[]{R.mipmap.ic_launcher, R.mipmap.ic_launcher, R.mipmap.ic_launcher};
@Override
protected void onCreate(Bundle savedInstanceState) {
    super.onCreate(savedInstanceState);
    setContentView(R.layout.listview_layout);
    ListView listView = (ListView)findViewById(R.id.listview1);
    //创建一个List集合,List集合的元素是MAP
    List<Map<String, Object>> list = new ArrayList<Map<String, Object>>();
    for(int i = 0;i < name.length;i++){
        Map<String, Object> listitem = new HashMap<String, Object>();
        listitem.put("icon", icon[i]);
        listitem.put("name", name[i]);
        listitem.put("desc", desc[i]);
        list.add(listitem);
    }
    //创建一个SimpleAdapter
    SimpleAdapter adapter = new SimpleAdapter(this, list, R.layout.list_item_layout, new String[]
{"name", "icon", "desc"}, new int[]{R.id.name, R.id.icon, R.id.dexc});
    listView.setAdapter(adapter);
    }
}
```

以上程序实现的效果如图 2-20 所示。

上面程序中加粗的代码是使用 SimpleAdapter 的重要一步。使用 SimpleAdapter 最重要的是它的 5 个参数,尤其是后面 4 个,下面介绍这 4 个参数。首先第二个参数是 List <? extends Map < String,? >>类型的集合对象,该集合中每个 Map < String,? >对象生成一行;第三个参数是指定一个界面布局的 ID,这里引用了一个自定义的布局 list_item_layout.xml 文件;第四个参数是 String 类型的参数,该参数决定提取哪些内容显示在 ListView 的每一行;最后一个是 int 类型的参数,决定显示哪些组件。

图 2-20  基于 SimpleAdapter 实现

**实例三:基于 BaseAdapter 创建 ListView**

在使用 SimpleAdapter 时,用户可以在布局当中定义按钮,但是当用户单击时,由于单击操作被 ListView 的 item(元素)所覆盖,导致按钮无法获取到焦点,这时候最方便的方法就是使用灵活的适配器 BaseAdapter。

BaseAdapter 是最基础的 Adapter,也就是说,它可以做所有的事情。说它最实用、最常用,原因就在于它的全能性,不会像 ArrayAdapter 等封装好的类有那么多局限性,但是这样的话,使用起来自然会更加麻烦。

使用 BaseAdapter 可以新建一个 Java 文件 MyBaseAdapter(继承自 BaseAdapter),并且重写它

的 4 个基础方法。代码如下所示：

```java
public class MyBaseAdapter extends BaseAdapter {
    @Override
    public int getCount() {
        return 0;
    }
    @Override
    public Object getItem(int position) {
        return null;
    }
    @Override
    public long getItemId(int position) {
        return 0;
    }
    @Override
    public View getView(int position, View convertView, ViewGroup parent) {
        return null;
    }
}
```

学会 BaseAdapter 其实只需要掌握 4 个方法：getCount()、getItem()、getItemId()和 getView()。关于每个方法的具体的含义如下所示：

- getCount()——要绑定的条目的数目,比如格子的数量。
- getItem()——根据一个索引(位置)获得该位置的对象。
- getItemId()——获取条目的 ID。
- getView()——获取该条目要显示的界面。

可以理解为 Adapter 先从 getCount()里确定数量,然后循环执行 getView()方法将条目一个一个绘制出来,所以必须重写的是 getCount()和 getView()方法。而 getItem()和 getItemId()是调用某些函数才会触发的方法,如果不需要使用可以暂时不修改。接下来将通过具体的例子来讲解 BaseAdapter 的使用。

首先创建 chapter3_BaseAdapter 项目,修改 activity_main.xml 的代码,添加一个 ListView 组件。具体如下所示：

```xml
<?xml version = "1.0" encoding = "utf-8"?>
<LinearLayout xmlns:android = "http://schemas.android.com/apk/res/android"
    android:layout_width = "match_parent"
    android:layout_height = "match_parent">
    <ListView
        android:id = "@ + id/listview"
        android:layout_width = "match_parent"
        android:layout_height = "wrap_content" />
</LinearLayout>
```

BaseAdapter 的使用与 SimpleAdapter 类似,都需要在 layout 目录下添加一个自定义的布局文件 list_item_layout.xml（即每一行的布局样式）。代码如下所示：

```xml
<?xml version = "1.0" encoding = "utf-8"?>
< LinearLayout xmlns:android = "http://schemas.android.com/apk/res/android"
    android:layout_width = "match_parent"
    android:layout_height = "match_parent"
    android:orientation = "horizontal" >
        < LinearLayout
            android:layout_width = "200dip"
            android:layout_height = "match_parent"
            android:orientation = "horizontal">
        < ImageView
            android:id = "@ + id/imageview"
            android:layout_width = "50dip"
            android:layout_height = "50dip" />
        < TextView
            android:id = "@ + id/textview"
            android:layout_width = "wrap_content"
            android:layout_height = "match_parent"
            android:paddingTop = "8dip"
            android:textSize = "20sp" />
    </LinearLayout >
    < Button
        android:id = "@ + id/button"
        android:layout_width = "wrap_content"
        android:layout_height = "wrap_content" />
</LinearLayout >
```

接着就是最重要的一步，即自定义一个 MyAdapter 类继承自 BaseAdapter，然后重写其中的方法。具体代码如下所示：

```java
import android.content.Context;
import android.view.LayoutInflater;
import android.view.View;
import android.view.ViewGroup;
import android.widget.BaseAdapter;
import android.widget.Button;
import android.widget.ImageView;
import android.widget.TextView;
import android.widget.Toast;
import java.util.List;
import java.util.Map;

public class MyAdapter extends BaseAdapter {
    private List < Map < String, Object >> datas;
    private Context mContext;
    /** 构造函数
     * datas 表示需要绑定到 View 的数据,mContext 表示传入上下文 */
```

```java
public MyAdapter(List<Map<String, Object>> datas, Context mContext) {
    this.datas = datas;
    this.mContext = mContext;
}
@Override
public int getCount() {
    //返回数据的总数
    return datas.size();
}
@Override
public Object getItem(int position) {
    //返回在List中指定位置的数据的内容
    return datas.get(position);
}
@Override
public long getItemId(int position) {
    //返回数据在List中所在的位置
    return position;
}
@Override
public View getView(int position, View convertView, ViewGroup parent) {
    final ViewHolder holder;
    if (convertView == null) {
        //使用自定义的布局文件作为Layout
        convertView = LayoutInflater.from(mContext).inflate(R.layout.list_item_layout, null);
        //减少findView的次数
        holder = new ViewHolder();
        //初始化布局中的元素
        holder.mImageView = (ImageView) convertView.findViewById(R.id.imageview);
        holder.mTextView = (TextView) convertView.findViewById(R.id.textview);
        holder.mButton = (Button) convertView.findViewById(R.id.button);
        holder.mButton.setOnClickListener(new View.OnClickListener() {
            @Override
            public void onClick(View view) {
                Toast.makeText(mContext,"你点了我!哈哈",Toast.LENGTH_SHORT).show();
            }
        });
        convertView.setTag(holder);
    } else {
        holder = (ViewHolder) convertView.getTag();
    }
    //从传入的数据中提取数据并绑定到指定的View中
    holder.mImageView.setImageResource((Integer) datas.get(position).get("img"));
    holder.mTextView.setText(datas.get(position).get("title").toString());
    holder.mButton.setText(datas.get(position).get("button").toString());
    return convertView;
}
static class ViewHolder {
```

```
        ImageView mImageView;
        TextView mTextView;
        Button mButton;
    }
}
```

最后就是在MainActivity中添加数据以及为ListView添加上文自定义的Adapter。具体代码如下所示:

```
import java.util.ArrayList;
import java.util.HashMap;
import java.util.List;
import java.util.Map;
import android.app.Activity;
import android.os.Bundle;
import android.widget.ListView;
import com.jxust.cn.chapter3_baseadapter.MyAdapter;

public class MainActivity extends Activity {
    private ListView mListView;
    private MyAdapter myAdapter;
    private List<Map<String, Object>> list = new ArrayList<Map<String, Object>>();
    @Override
    protected void onCreate(Bundle savedInstanceState) {
        super.onCreate(savedInstanceState);
        setContentView(R.layout.activity_main);
        initData();
        mListView = (ListView) findViewById(R.id.listview);
        myAdapter = new MyAdapter(list, this);
        mListView.setAdapter(myAdapter);
    }
    //自定义数据,也可以添加网络数据
    private void initData() {
        Map<String, Object> map = new HashMap<String, Object>();
        map.put("img", R.drawable.android);
        map.put("title", "Android");
        map.put("button", "学习");
        list.add(map);
        map = new HashMap<String, Object>();
        map.put("img", R.drawable.java1);
        map.put("title", "Java");
        map.put("button", "学习");
        list.add(map);
        map = new HashMap<String, Object>();
        map.put("img", R.drawable.html5);
        map.put("title", "HTML5");
        map.put("button", "学习");
```

```
            list.add(map);
            map = new HashMap<String, Object>();
            map.put("img", R.drawable.c1);
            map.put("title", "C");
            map.put("button", "学习");
            list.add(map);
            map = new HashMap<String, Object>();
            map.put("img", R.drawable.python);
            map.put("title", "Python");
            map.put("button", "学习");
            list.add(map);
        }
    }
```

运行程序,单击界面上的自定义按钮,实现的效果如图 2-21 所示。

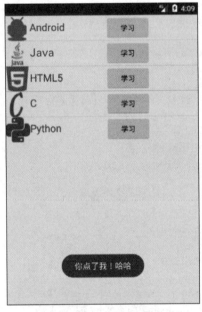

图 2-21　基于 BaseAdapter 实现

视频讲解

## 2.5　对话框的使用

在 Android 开发中,经常需要在 Android 界面上弹出一些对话框,询问用户或者让用户选择。实现这些功能的组件称为 Android 对话框(Dialog),本节将通过一个案例讲解对话框 AlertDialog 的使用。

AlertDialog 对话框的功能很强大,使用它可以生成各种有内容的对话框,使用 AlertDialog 对话框主要有以下几个步骤:

- 创建 AlertDialog.Builder 对象。

- 创建 AlertDialog.Builder.setTitle()或 setCustomTitle()方法设置标题。
- 调用 AlertDialog.Builder.setIcon()设置对话框图标。
- 调用 AlertDialog.Builder.setPositiveButton()等添加按钮。
- 调用 AlertDialog.Builder 的 create()方法创建 AlertDialog 对象,再调用 AlertDialog 的 show()方法把对话框显示出来。

接下来将通过具体的例子讲解 AlertDialog 的使用。

布局文件采用了线性布局的方式,在布局中添加一个 Button 组件,然后添加单击事件,单击以后出现一个对话框,具体的实现代码如下：

```xml
<?xml version = "1.0" encoding = "utf-8"?>
< LinearLayout xmlns:android = "http://schemas.android.com/apk/res/android"
    android:layout_width = "match_parent"
    android:layout_height = "match_parent">
    < Button
        android:id = "@ + id/dialog"
        android:layout_width = "wrap_content"
        android:layout_height = "wrap_content"
        android:layout_weight = "1"
        android:layout_marginTop = "200dp"
        android:text = "显示对话框" />
</LinearLayout >
```

然后在 Activity 中初始化 Button,为 Button 添加单击事件,创建对话框,具体代码如下所示：

```java
public class MainActivity extends Activity {
@Override
    protected void onCreate(Bundle savedInstanceState) {
        super.onCreate(savedInstanceState);
        setContentView(R.layout.dialog_layout);
        //对话框的使用
        Button button = (Button)findViewById(R.id.dialog);
        button.setOnClickListener(new View.OnClickListener() {
            @Override
            public void onClick(View view) {
            //设置对话框标题
            new AlertDialog.Builder(MainActivity.this).setTitle("系统提示")
            //设置显示的内容
            .setMessage("请确认所有数据都保存后再推出系统!")
            //添加"确定"按钮
            .setPositiveButton("确定",new DialogInterface.OnClickListener() {
                public void onClick(DialogInterface dialog, int which) {finish();
                }
        }).setNegativeButton("返回",new DialogInterface.OnClickListener() {     //添加"返回"按钮
                    @Override
                    public void onClick(DialogInterface dialog, int which) {
```

```
            }
        }).show();                    //在按键响应事件中显示此对话框
        }
    });
    }
}
```

上述代码创建的对话框如图 2-22 所示。

图 2-22  对话框 AlertDialog

从上面的代码中可以看到创建一个对话框基本的步骤。这只是一个基本的对话框,关于其他类型的对话框,例如单选对话框、多选对话框以及自定义视图对话框,在后续章节中会结合具体的项目讲解。

## 2.6  Toast 的使用

在 Android 的实际开发过程中,经常会使用 Toast 作为调试工具,通过 Toast 组件显示传递的变量值等,观察是否跟预想情况一样。

Toast 会显示一个消息在屏幕上,告诉用户一些信息,并且在短暂的时间后会自动消失。使用 Toast 需要掌握两个方法,分别是 makeText()方法和 show()方法。makeText()方法用于设置要提示用户的文字,包含 3 个参数,分别为组件的上下文环境、要显示的文字、显示时间的长短。显示时间的长短通常使用 Toast.LENGTH_SHORT 和 Toast.LENGTH_LONG 表示,也可以使用 0 和 1 分别代表 SHORT 和 LONG。

使用 Toast,只需要在 Activity 中添加如下代码:

```
Toast.makeText(this,"要显示的文字",Toast.LENGTH_SHORT).show();
```

运行效果如图 2-23 所示。

图 2-23　Toast 组件

## 2.7　Log 的使用

Log 是 Android 提供的用于输出日志的工具类。每个 Android 的日志消息都有与其关联的标志和不同的优先级，用来表示不同的信息。

- Log.v()——verbose，打印所有日志信息，优先级最低。
- Log.d()——debug，打印调试信息，优先级高于 verbose。
- Log.i()——info，打印用户需要的信息，优先级高于 debug。
- Log.w()——warn，打印警告信息，优先级高于 info。
- Log.e()——error，打印程序中的错误信息，优先级高于 warn。

上述 Log 的方法中包含两个参数，如 Log.i(tag,msg)，tag 表示标签，msg 表示要打印的信息。当使用 Log 打印日志时，可以通过 Logcat 来查看输出的日志。Logcat 是 Android 提供的一个命令行工具，用于转储消息日志。下面介绍使用 Logcat 查看自定义 Log 的方法。

创建新项目后，在编辑窗口下方可以看到 Logcat 图标，如图 2-24 所示。

单击 Logcat，进入系统运行的日志界面，单击 Regex 的下拉列表框，选择 Edit Filter Configuration，如图 2-25 所示。

图 2-24　选择 Logcat

图 2-25　选择自定义 Log

之后弹出 Create New Logcat Filter 对话框，在 Log Level 下可以选择 Log 的级别，默认是 Verbose。也可以在 Log Tag 处输入自定义的 tag，如果一个项目中有多个不同的 tag，可以在 Package Name 处输入项目的包名。

例如在 MainActivity 的 onCreate()方法中加入语句 Log.i("Activity","start")，则 Logcat 的配置如图 2-26 所示。运行 MainActivity，其 Log 输出日志显示如图 2-27 所示。

图 2-26　编辑 Log 的过滤条件

图 2-27　Logcat 查看 Log 输出日志

## 2.8　用户注册案例布局讲解

前几节分别介绍了 Android 的布局方式以及 Android 中组件在 XML 文件中的定义使用，接下来将通过一个用户注册的综合案例讲解各个组件的组合使用。

首先应该明确用户注册需要哪些信息以及显示这些信息所对应的组件，根据分析应该展示如下信息。

- 手机号：需要输入手机号信息，所以使用 EditText 组件，以及用来显示手机号的 TextView 组件。
- 密码：用户需要输入密码，所以使用 EditText 组件显示密码的 TextView 组件。
- 性别：选择性别，使用一组 RadioGroup，其中包含两个 RadionButton。
- 兴趣爱好：用户可能有多个爱好，需要使用复选框组件。
- 地址：用户选择所在城市，所以使用 Spinner 下拉列表框组件。
- "注册"按钮：Button 组件，单击该按钮进行注册。

结合以上信息，可知整体布局方式采用一种较为常用的线性垂直布局，在 TextView 和 EditText 组合使用时，采用线性水平的布局方式。具体的设计代码如下所示：

```
<?xml version = "1.0" encoding = "utf-8"?>
< LinearLayout xmlns:android = "http://schemas.android.com/apk/res/android"
    xmlns:app = "http://schemas.android.com/apk/res - auto"
```

```xml
    xmlns:tools = "http://schemas.android.com/tools"
    android:layout_width = "match_parent"
    android:layout_height = "match_parent"
    android:orientation = "vertical"
    tools:context = "com.jxust.cn.chapter2_zonghe.MainActivity">
    <!-- 手机号 -->
    <LinearLayout
        android:layout_width = "368dp"
        android:layout_height = "wrap_content"
        tools:layout_editor_absoluteY = "0dp"
        android:orientation = "horizontal"
        tools:layout_editor_absoluteX = "8dp">
        <TextView
            android:layout_width = "wrap_content"
            android:layout_height = "40dp"
            android:textSize = "18sp"
            android:textColor = "@android:color/background_dark"
            android:text = "手机号:"/>
        <EditText
            android:layout_width = "match_parent"
            android:layout_height = "50dp"
            android:hint = "请输入手机号"/>
    </LinearLayout>
    <!-- 密码 -->
    <LinearLayout
        android:layout_width = "368dp"
        android:layout_height = "wrap_content"
        tools:layout_editor_absoluteY = "0dp"
        android:orientation = "horizontal"
        tools:layout_editor_absoluteX = "8dp">
        <TextView
            android:layout_width = "wrap_content"
            android:layout_height = "40dp"
            android:textSize = "18sp"
            android:textColor = "@android:color/background_dark"
            android:text = "密    码:"/>
        <EditText
            android:layout_width = "match_parent"
            android:layout_height = "50dp"
            android:hint = "请输入密码"/>
    </LinearLayout>
    <!-- 性别选择 -->
        <RadioGroup
            android:layout_width = "match_parent"
            android:layout_height = "40dp"
            android:orientation = "horizontal">
            <RadioButton
                android:layout_width = "50dp"
```

```xml
            android:layout_height = "wrap_content"
            android:checked = "true"
            android:text = "男"/>
        < RadioButton
            android:layout_width = "50dp"
            android:layout_height = "wrap_content"
            android:text = "女"/>
    </RadioGroup>
    < LinearLayout
        android:layout_width = "match_parent"
        android:layout_height = "wrap_content"
        android:orientation = "horizontal">
        < CheckBox
            android:layout_width = "wrap_content"
            android:layout_height = "wrap_content"
            android:text = "读书" />
        < CheckBox
            android:layout_width = "wrap_content"
            android:layout_height = "wrap_content"
            android:checked = "true"
            android:text = "打球" />
        < CheckBox
            android:layout_width = "wrap_content"
            android:layout_height = "wrap_content"
            android:text = "听音乐" />
    </LinearLayout>
    < Spinner
        android:id = "@ + id/spinner"
        android:layout_width = "match_parent"
        android:layout_height = "wrap_content" />
    < Button
        android:layout_width = "fill_parent"
        android:background = "#3F51B5"
        android:textColor = "#FFFFFF"
        android:textSize = "18sp"
        android:text = "注 册"
        android:layout_height = "40dp" />
</LinearLayout>
```

因为使用到了 Spinner 下拉列表框组件,所以在 Activity 中使用 ArrayAdapter 为其添加数据。Activity 中添加数据代码如下:

```
Spinner spinner = (Spinner)findViewById(R.id.spinner);
String[] city = new String[]{"北京","上海","武汉","南京","南昌"};
ArrayAdapter < String > adapter = new ArrayAdapter < String >(this,
android.R.layout.simple_list_item_1,city);
spinner.setAdapter(adapter);
```

上述代码实现的用户注册界面如图 2-28 所示。

图 2-28　用户注册界面

## 2.9 【项目案例 2】 "橙一色"App 注册/登录界面设计

视频讲解

1.4 节中创建了"橙一色"App 项目,在学习完本章 Android 开发常用的界面组件以及布局以后,为"橙一色"App 添加注册和登录的界面。

### 1. 登录界面设计

登录界面主要使用 EditText、Button 以及 ImageView 组件,通过设置组件的属性以及设置组件的样式来完成。首先在 layout 下新建 user_login.xml 文件,具体的内容如下所示:

```xml
<?xml version = "1.0" encoding = "utf-8"?>
< LinearLayout xmlns:android = "http://schemas.android.com/apk/res/android"
    android:layout_width = "match_parent"
    android:layout_height = "match_parent"
    android:orientation = "vertical"
    android:background = "♯E8E8E8">
    <!-- App Logo 及名字 -->
    < LinearLayout
        android:layout_width = "match_parent"
        android:layout_height = "wrap_content"
        android:gravity = "center_horizontal"
        android:layout_marginTop = "40dp"
        android:orientation = "vertical">
        < ImageView
            android:layout_width = "wrap_content"
            android:layout_height = "wrap_content"
            android:layout_gravity = "center_horizontal"
```

```xml
            android:src = "@drawable/orange" />
        <TextView
            android:layout_width = "wrap_content"
            android:layout_height = "wrap_content"
            android:textSize = "22sp"
            android:textColor = "#000000"
            android:textStyle = "italic"
            android:text = "Orange" />
    </LinearLayout>
    <LinearLayout
        android:layout_width = "match_parent"
        android:layout_height = "wrap_content"
        android:gravity = "center_horizontal"
        android:layout_marginTop = "50dp"
        android:orientation = "horizontal">
        <ImageView
            android:layout_width = "30dp"
            android:layout_height = "30dp"
            android:layout_gravity = "center_vertical"
            android:src = "@drawable/user" />
        <EditText
            android:id = "@+id/username"
            android:layout_width = "270dp"
            android:layout_height = "60dp"
            android:textAlignment = "center"
            android:maxLength = "10"
            android:maxLines = "1"
            android:hint = "username" />
    </LinearLayout>
    <LinearLayout
        android:layout_width = "match_parent"
        android:layout_height = "wrap_content"
        android:gravity = "center_horizontal"
        android:orientation = "horizontal">
        <ImageView
            android:layout_width = "30dp"
            android:layout_height = "30dp"
            android:layout_gravity = "center_vertical"
            android:src = "@drawable/password" />
        <EditText
            android:id = "@+id/password"
            android:layout_width = "270dp"
            android:layout_height = "60dp"
            android:inputType = "textPassword"
            android:textAlignment = "center"
            android:maxLength = "16"
```

```xml
            android:maxLines = "1"
            android:hint = "password" />
    </LinearLayout>
    <Button
        android:id = "@+id/login"
        android:layout_width = "250dp"
        android:layout_height = "50dp"
        android:textSize = "18sp"
        android:layout_gravity = "center_horizontal"
        android:layout_marginTop = "60dp"
        android:background = "@drawable/button_login"
        android:textColor = "#FFFFFF"
        android:text = "登录" />
    <Button
        android:id = "@+id/register"
        android:layout_width = "250dp"
        android:layout_height = "50dp"
        android:layout_marginTop = "10dp"
        android:textSize = "18sp"
        android:layout_gravity = "center_horizontal"
        android:background = "@drawable/button_register"
        android:textColor = "#FFFFFF"
        android:text = "注册" />
</LinearLayout>
```

为了使设计出来的界面更好看,采用自定义样式的方式,自定义样式是通过在 drawable 下添加图片以及创建样式.xml 文件,本项目中为 Button 组件自定义了样式,用来区分单击和松开的不同操作。创建 button_login.xml 如下所示:

```xml
<?xml version = "1.0" encoding = "utf-8"?>
<selector xmlns:android = "http://schemas.android.com/apk/res/android">
    <item android:state_pressed = "false">
        <shape>
            <solid android:color = "#EFB81C" />
            <corners android:radius = "5dp" />
        </shape>
    </item>
    <item android:state_pressed = "true">
        <shape>
            <solid android:color = "#c0c0c0" />
            <corners android:radius = "5dp" />
        </shape>
    </item>
</selector>
```

采用一个 XML 来描述状态列表,通过唯一的 selector 节点,利用 item 来描述每一种状态。item 的属性包括:

- android:drawable：应用一个 drawable 资源；
- android:state_focused：表示对象获取到焦点；
- android:state_pressed：对象被按压；
- android:state_hovered：经过对象时触发；
- android:state_selected：对象被选中。

Button 组件的按下效果是通过<shape>标签来完成的。<shape>标签用来指定背景的样式，shape 的常用属性：stroke，用来设定背景边框的样式，可以定义边框的宽度（width）、颜色（color）、是否为虚线展示等；solid，用来设定背景颜色；corners，表示边框 4 个角的弧度。

最终"橙一色"App 的登录界面如图 2-29 所示。

### 2. 注册界面设计

"橙一色"App 的注册界面主要包含了用于输入账户名、密码和确认密码的 EditText 组件，用于选择性别的 RadioGroup 和 RadioButton 组件，还有用于显示所选择地址的 TextView 组件。地址的选择使用了第三方的组件 CityPicker 省市区联动插件，便于快速选择地区。注册界面布局的具体内容如下：

图 2-29 "橙一色"登录界面

```
<?xml version = "1.0" encoding = "utf-8"?>
<LinearLayout xmlns:android = "http://schemas.android.com/apk/res/android"
...
    android:background = "#E8E8E8">
    <!-- App Logo 及名字 -->
    <LinearLayout
        android:layout_width = "match_parent"
        android:layout_height = "wrap_content"
        android:gravity = "center_horizontal"
        android:layout_marginTop = "20dp"
        android:orientation = "vertical">
        <ImageView
            android:layout_width = "wrap_content"
            android:layout_height = "wrap_content"
            android:layout_gravity = "center_horizontal"
            android:src = "@drawable/orange" />
        <TextView
            android:layout_width = "wrap_content"
            android:layout_height = "wrap_content"
            android:textSize = "22sp"
            android:textColor = "#000000"
            android:textStyle = "italic"
            android:text = "Orange" />
    </LinearLayout>
```

```xml
<LinearLayout
    android:layout_width = "match_parent"
    android:layout_height = "wrap_content"
    android:gravity = "center_horizontal"
    android:layout_marginTop = "30dp"
    android:orientation = "horizontal">
    <ImageView
        android:layout_width = "30dp"
        android:layout_height = "30dp"
        android:layout_gravity = "center_vertical"
        android:src = "@drawable/user" />
    <EditText
        android:id = "@ + id/reg_username"
        android:layout_width = "290dp"
        android:layout_height = "60dp"
        android:textAlignment = "center"
        android:maxLength = "10"
        android:maxLines = "1"
        android:hint = "用户名10位" />
</LinearLayout>
<LinearLayout
    android:layout_width = "match_parent"
    android:layout_height = "wrap_content"
    android:gravity = "center_horizontal"
    android:orientation = "horizontal">
    <ImageView
        android:layout_width = "30dp"
        android:layout_height = "30dp"
        android:layout_gravity = "center_vertical"
        android:src = "@drawable/password" />
    <EditText
        android:id = "@ + id/reg_password"
        android:layout_width = "290dp"
        android:layout_height = "60dp"
        android:inputType = "textPassword"
        android:textAlignment = "center"
        android:maxLength = "16"
        android:maxLines = "1"
        android:hint = "密码6－16位" />
</LinearLayout>
<LinearLayout
    android:layout_width = "match_parent"
    android:layout_height = "wrap_content"
    android:gravity = "center_horizontal"
    android:orientation = "horizontal">
    <ImageView
```

```xml
            android:layout_width = "30dp"
            android:layout_height = "30dp"
            android:layout_gravity = "center_vertical"
            android:src = "@drawable/sure_apssword" />
        <EditText
            android:id = "@ + id/reg_sure_password"
            android:layout_width = "290dp"
            android:layout_height = "60dp"
            android:inputType = "textPassword"
            android:textAlignment = "center"
            android:maxLength = "16"
            android:maxLines = "1"
            android:hint = "确认密码" />
</LinearLayout>
<LinearLayout
        android:layout_width = "match_parent"
        android:layout_height = "wrap_content"
        android:gravity = "center_horizontal"
        android:layout_marginTop = "10dp"
        android:orientation = "horizontal">
        <ImageView
            android:layout_width = "30dp"
            android:layout_height = "30dp"
            android:layout_gravity = "center_vertical"
            android:src = "@drawable/sex" />
        <RadioGroup
            android:layout_width = "270dp"
            android:layout_height = "wrap_content"
            android:orientation = "horizontal">
            <RadioButton
                android:id = "@ + id/reg_man"
                android:layout_width = "wrap_content"
                android:checked = "true"
                android:layout_height = "wrap_content"
                android:text = "男"/>
            <RadioButton
                android:id = "@ + id/reg_woman"
                android:layout_width = "wrap_content"
                android:layout_height = "wrap_content"
                android:text = "女"/>
        </RadioGroup>
</LinearLayout>
<LinearLayout
        android:layout_width = "match_parent"
        android:layout_height = "wrap_content"
        android:gravity = "center_horizontal"
```

```xml
            android:layout_marginTop = "10dp"
            android:orientation = "horizontal">
            <ImageView
                android:layout_width = "30dp"
                android:layout_height = "30dp"
                android:layout_gravity = "center_vertical"
                android:src = "@drawable/address" />
            <TextView
                android:id = "@+id/reg_province"
                android:layout_width = "290dp"
                android:layout_height = "60dp"
                android:gravity = "center"
                android:maxLines = "1"
                android:drawableRight = "@drawable/arrow_down"
                android:hint = "地址选择" />
        </LinearLayout>
        <Button
            android:id = "@+id/reg_register"
            android:layout_width = "280dp"
            android:layout_height = "50dp"
            android:layout_marginTop = "80dp"
            android:textSize = "18sp"
            android:layout_gravity = "center_horizontal"
            android:background = "@drawable/button_login"
            android:textColor = "#FFFFFF"
            android:text = "注册" />
</LinearLayout>
```

其中使用 CityPicker 时需要在 RegisterActivity 中进行初始化,初始化的内容主要包括默认的地址、标题和背景颜色等属性。其中 RegisterActivity 的具体内容如下所示:

```java
public class RegisterActivity extends AppCompatActivity implements View.OnClickListener {
    private TextView cityText;
    private CityPicker cityPicker;
    @Override
    protected void onCreate(@Nullable Bundle savedInstanceState) {
        super.onCreate(savedInstanceState);
        setContentView(R.layout.user_register);
        init();
    }
    //界面组件初始化
    private void init() {
        cityText = findViewById(R.id.reg_province);
        cityText.setOnClickListener(this);
    }
    @Override
```

```java
public void onClick(View v) {
    switch (v.getId()) {
        case R.id.reg_province:
            initCityPicker();
            cityPicker.show();
            break;
    }
}
//初始化城市选择器
public void initCityPicker() {
    cityPicker = new CityPicker.Builder(RegisterActivity.this)
            .textSize(16)
            .title("地址选择")
            .backgroundPop(0xa0000000)
            .titleBackgroundColor("#EFB81C")
            .titleTextColor("#000000")
            .backgroundPop(0xa0000000)
            .confirTextColor("#000000")
            .cancelTextColor("#000000")
            .province("xx省")
            .city("xx市")
            .district("xx区")
            .textColor(Color.parseColor("#000000"))
            .provinceCyclic(true)
            .cityCyclic(false)
            .districtCyclic(false)
            .visibleItemsCount(7)
            .itemPadding(10)
            .onlyShowProvinceAndCity(false)
            .build();
    cityPicker.setOnCityItemClickListener(new CityPicker.OnCityItemClickListener() {
        @Override
        public void onSelected(String...strings) {
            String province = strings[0];
            String city = strings[1];
            String district = strings[2];
            cityText.setText(String.format("%s %s %s", province, city, district));
        }
        @Override
        public void onCancel() {
        }
    });
}
```

上述代码中，加粗的initCityPicker()方法即是城市选择器的初始化方法，其中省、市、县、区和邮政编码都存储在strings数组当中，如上加粗代码，分别取出了省、市和区。

图 2-30　注册界面

## 2.10　本章小结

本章主要介绍了 Android 开发的 7 个布局方式以及 UI 界面设计中所使用到的组件,这些基本的布局方式以及组件的使用需要熟练掌握。学会组件的自由搭配以及自定义样式才能更好地进行 Android 程序的开发。"橙一色"App 登录和注册界面的设计,融合了自定义样式及第三方插件的使用,在项目开发中可以借鉴使用。

## 2.11　课后习题

1. 分别用 Android 的 7 种布局方式来设计一个界面。
2. 说明 TextView 和 EditText 的关系。
3. 使用 ArrayAdapter 的方式实现一个城市选择的下拉列表例子。
4. 设计一个登录界面,包含"用户名"和"密码"两个文本输入框、"记住密码"复选框、"忘记密码"按钮、"登录"按钮,忘记密码时需要弹出对话框。

# 第 3 章

# Activity

**学习目标**
- 掌握 Activity 的生命周期。
- 掌握 Activity 的常用方法。
- 掌握显式 Intent 和隐式 Intent 的使用。
- 掌握 Activity 中的数据传递方式。

在 Android 系统中,用户与程序的交互是通过 Activity 完成的,同时 Activity 也是 Android 四大组件中使用最多的一个,本章将详细讲解有关 Activity 的知识。

## 3.1 Activity 基础

### 3.1.1 认识 Activity

Activity 的中文意思是"活动",它是 Android 应用中负责与用户交互的组件,相当于 Swing 编程中的 JFrame 组件。与其不同的是,JFrame 本身可以设置布局管理器,不断地向其添加组件,而 Activity 只能通过 setContentView(View)来显示布局文件中已经定义的组件。

在应用程序中,Activity 就像一个界面管理员,用户在界面上的操作是通过 Activity 来管理的,下面是 Activity 的常用事件。

- OnKeyDown(int keyCode,KeyEvent event):按键按下事件。
- OnKeyUp(int keyCode,KeyEvent event):按键松开事件。
- OnTouchEvent(MotionEvent event):单击屏幕事件。

当用户按下手机界面上的按键时,就会触发 Activity 中对应的事件 OnKeyDown()来响应用户的操作。3.1.2 节会通过具体实例讲解 Activity 的常用事件。

## 3.1.2 如何创建 Activity

创建一个 Activity 的具体的步骤如下：

（1）定义一个类继承自 android.app.Activity 或其子类，如图 3-1 所示。

图 3-1 创建 Activity

（2）在 res/layout 目录下创建一个 XML 文件，用于创建 Activity 的布局。

（3）在 app/manifests 目录下的 AndroidManifest.xml 清单文件中注册 Activity，如图 3-2 所示。

图 3-2 Activity 注册

（4）重写 Activity 的 onCreate()方法，并在该方法中使用 setContentView()加载指定的布局文件。新创建的 Activity 的具体代码如下所示：

```
package com.jxust.cn.chapter3_1;
import android.app.Activity;
import android.os.Bundle;
public class MainActivity extends Activity {
//项目的入口 Activity
    @Override
    protected void onCreate(Bundle savedInstanceState) {
        super.onCreate(savedInstanceState);
        //设置 Activity 显示的布局
        setContentView(R.layout.activity_main);
    }
}
```

接下来将通过具体的例子讲解 3.1.1 节中提到的 Activity 中的几个常用事件，具体的 Activity 代码如下所示：

```
public class MainActivity extends Activity {
    @Override
    protected void onCreate(Bundle savedInstanceState) {
```

```
        super.onCreate(savedInstanceState);
        setContentView(R.layout.activity_main);
}
//响应按键按下事件
public boolean onKeyDown(int keycode, KeyEvent event){
Toast.makeText(this,"按键按下了!", Toast.LENGTH_SHORT).show();
    return super.onKeyDown(keycode, event);
}
//响应按键松开事件
public boolean onKeyUp(int keycode,  KeyEvent event){
  Toast.makeText(this,"按键松开了!", Toast.LENGTH_SHORT).show();
    return super.onKeyDown(keycode, event);
}
//响应屏幕触摸操作
public boolean onTouchEvent( MotionEvent event){
  Toast.makeText(this,"触摸了屏幕!", Toast.LENGTH_SHORT).show();
    return super.onTouchEvent(event);
 }
}
```

运行效果如图 3-3 所示。

图 3-3　Activity 常用事件

视频讲解

## 3.1.3　Activity 的生命周期

每一个 Android 应用程序在运行时,对于底层的 Linux Kernel 而言都是一个单独的进程,但是对于 Android 系统而言,因为局限于手机画面的大小与使用的考虑,不能把每一个运行中的应用程序窗口都显示出来。所以通常手机系统的界面一次仅显示一个应用程序窗口，Android 使用了 Activity 的概念来表示界面。

Activity 的生命周期分为 3 种状态，分别是运行状态、暂停状态和停止状态。下面将详细介绍这 3 种状态。

- 运行状态：当 Activity 在屏幕最前端的时候，它是有焦点的、可见的，可以供用户进行单击、长按等操作，这种状态称为运行状态。
- 暂停状态：在一些情况下，最上层的 Activity 没有完全覆盖屏幕，这时候被覆盖的 Activity 仍然对用户可见，并且存活。但当内存不足时，这个暂停状态的 Activity 可能会被杀死。
- 停止状态：当 Activity 完全不可见时，它就处于了停止状态，但仍然保留着当前状态和成员信息，当系统内存不足时，这个 Activity 就很容易被杀死。

Activity 从一种状态变到另一种状态时会经过一系列 Activity 类的方法。常用的回调方法如下：

- onCreate(Bundle savedInstanceState)——该方法在 Activity 的实例被 Android 系统创建后第一个被调用。通常在该方法中设置显示屏幕的布局、初始化数据、设置组件被单击的事件响应代码。
- onStart()——在 Activity 可见时执行。
- onReStart()——回到最上边的界面，再次可见时执行。
- onResume()——Activity 获取焦点时执行。
- onPause()——Activity 失去焦点时执行。
- onStop()——用户不可见进入后台时执行。
- onDestroy()——Activity 销毁时执行。

为了更好地理解 Activity 的生命周期以及在 Activity 不同状态切换时所调用的方法，接下来将通过 Google 公司提供的一个 Activity 生命周期图来更生动地展示，如图 3-4 所示。

从图 3-4 中可以看出，Activity 在从启动到关闭的过程中，会依次执行 onCreate()→onStart()→onResume()→onPause()→onStop()→onDestroy()方法。如果进程被杀死，会重新执行 onCreate()方法。

为了更好地掌握 Activity 的生命周期中方法的执行过程，接下来将通过具体的例子来展现方法的执行顺序。

先创建一个布局界面，其中包含一个按钮，用来跳转到另一个 Activity 中使用，布局代码如下所示：

```
<?xml version = "1.0" encoding = "utf-8"?>
< LinearLayout xmlns:android = "http://schemas.android.com/apk/res/android"
    xmlns:app = "http://schemas.android.com/apk/res-auto"
    xmlns:tools = "http://schemas.android.com/tools"
    android:layout_width = "match_parent"
    android:layout_height = "match_parent"
    tools:context = "com.jxust.cn.chapter_shengtime.MainActivity">
    < Button
```

```
        android:id = "@ + id/button"
        android:layout_width = "wrap_content"
        android:layout_height = "wrap_content"
        android:layout_weight = "1"
        android:text = "跳转到第二个 Activity" />
</LinearLayout>
```

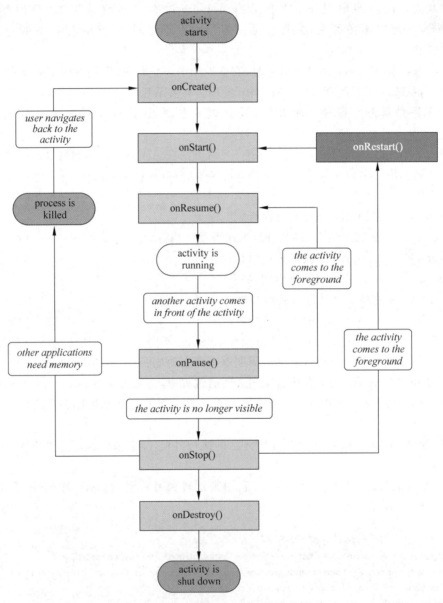

图3-4 Activity 的生命周期图

第一个 Activity 的代码如下所示：

```java
public class MainActivity extends Activity {
    //Activity1 创建时调用的方法
    @Override
    protected void onCreate(Bundle savedInstanceState) {
        super.onCreate(savedInstanceState);
        setContentView(R.layout.activity_main);
        Log.i("Activity1", "onCreate()");
        Button button = (Button)findViewById(R.id.button);
        button.setOnClickListener(new View.OnClickListener() {
            @Override
            public void onClick(View view) {
                Intent intent = new Intent(MainActivity.this, SecondActivity.class);
                startActivity(intent);
            }});
    }
    //Activity1 可见时调用的方法
    @Override
    protected void onStart() {
        super.onStart();
        Log.i("Activity1", "onStart()");
    }
    @Override
    protected void onRestart() {
        super.onRestart();
        Log.i("Activity1", "onRestart()");
    }
    //Activity1 获取到焦点时调用的方法
    @Override
    protected void onResume() {
        super.onResume();
        Log.i("Activity1", "onResume()");
    }
    //Activity1 失去焦点时调用的方法
    @Override
    protected void onPause() {
        super.onPause();
        Log.i("Activity1", "onPause()");
    }
    //Activity1 不可见时调用的方法
    @Override
    protected void onStop() {
        super.onStop();
        Log.i("Activity1", "onStop()");
    }
    //Activity1 被销毁时调用的方法
    @Override
```

```
protected void onDestroy() {
    super.onDestroy();
    Log.i("Activity1", "onDestroy()");
}
}
```

图 3-5　Activity1 界面

第二个 Activity 中的代码和使用的布局代码与第一个类似。

上面代码写好以后,在 AndroidManifest.xml 中注册创建的 Activity。完成上述步骤以后,运行界面如图 3-5 所示。

上述代码中使用了 Log 来打印日志信息。第二个 Activity 中的 Log.i 的 tag 是 Activity2,因而 Logcat 中选择设置 package name 为当前项目的包名。打印的 Activity 生命周期的执行方法顺序如图 3-6 所示。

```
2020-01-31 23:07:32.079 28759-28759/com.jxust.cn.chapter_shengtime I/Activity1: onCreate()
2020-01-31 23:07:32.093 28759-28759/com.jxust.cn.chapter_shengtime I/Activity1: onStart()
2020-01-31 23:07:32.095 28759-28759/com.jxust.cn.chapter_shengtime I/Activity1: onResume()
```

图 3-6　Activity1 的生命周期

从图 3-6 中的日志信息可以看出启动 Activity1 依次执行了 onCreate()、onStart()、onResume()方法,这是 Activity 从创建到可供用户操作的过程。

接下来单击第一个布局界面的按钮跳转到第二个 Activity,出现如图 3-7 所示的界面。

图 3-7　Activity2 界面

与此同时,Log 打印的日志信息如图 3-8 所示。

从图 3-8 的日志信息可以看出,当跳转到 Activity2 的时候,Activity1 会失去焦点,然后执行 onPause()方法,此时 Activity2 创建会一次执行 onCreate()、onStart()、onResume()方法。这时候 Activity1 会执行 onStop()方法。

```
2020-01-31 23:08:54.070 28759-28759/com.jxust.cn.chapter_shengtime I/Activity1: onPause()
2020-01-31 23:08:54.139 28759-28759/com.jxust.cn.chapter_shengtime I/Activity2: onCreate()
2020-01-31 23:08:54.152 28759-28759/com.jxust.cn.chapter_shengtime I/Activity2: onStart()
2020-01-31 23:08:54.156 28759-28759/com.jxust.cn.chapter_shengtime I/Activity2: onResume()
2020-01-31 23:08:54.901 28759-28759/com.jxust.cn.chapter_shengtime I/Activity1: onStop()
```

图 3-8　Activity1 跳转到 Activity2 的生命周期

接下来从第二个界面返回到第一个界面,此时 Log 打印的日志信息如图 3-9 所示。

从图 3-9 中的日志信息可以看出,从 Activity2 再次返回到 Activity1 时,Activity2 会先执行 onPause()方法,然后 Activity1 会依次执行 onRestart()方法、onStart()方法、onResume()方法,随后 Activity2 执行 onStop()方法和 onDestroy()方法。如果退出应用程序,则 Activity1 会执行 onStop()方法,然后执行 onDestroy()方法。

图 3-9　Activity2 跳转到 Activity1 的生命周期

### 3.1.4　Activity 中的单击事件

视频讲解

3.1.3 节学习了 Activity 中的生命周期，可以看到，从第一个界面到第二个界面使用了按钮的单击事件。在 Android 中 View 的单击事件共有 4 种，接下来详细讲解这 4 种方式。

第一种为在布局文件中设置按钮 onClick 属性为其指定 Activity 中的方法，代码如下所示：

```
<!-- 布局文件中添加单击事件为其指定方法名 -->
android:onClick = "click"
Activity 中的方法：
    public void click(View view){
        Intent intent = new Intent(MainActivity.this,SecondActivity.class);
        startActivity(intent);
    }
```

第二种是创建内部类的方式，创建一个内部类实现 OnClickListener 接口并重写 onClick() 方法，在方法中写入单击事件的逻辑。这一种方法不常用，所以不做详细介绍。

第三种就是主类实现 OnClickListener 接口，然后重写 onClick() 方法，并通过 switch() 语句判断哪个按钮被单击。具体的代码如下所示：

```
public class MainActivity extends Activity implements View.OnClickListener {
    protected void onCreate(Bundle savedInstanceState) {
        super.onCreate(savedInstanceState);
        setContentView(R.layout.activity_main);
        register = (Button)findViewById(R.id.register);
        register.setOnClickListener(this);
        @Override
        public void onClick(View view) {
            switch (view.getId())  {
                case R.id.register:
                    break;
            }
        }
    }
}
```

这里需要注意的是"register.setOnClickListener(this);"方法中这个 this 代表的是该 Activity 的引用，由于 Activity 实现了 OnClickListener 接口，所以这里就代表了 OnClickListener 的引用，在

方法中传入 this，就代表该组件绑定了单击事件的接口。

第四种就是匿名内部类的方式，适合按钮比较少的情况下使用。这种方式可以直接创建 OnClickListener 的匿名内部类传入按钮的 setOnClickListener()方法和参数中。具体的代码如下所示：

```java
public class MainActivity extends Activity {
    //Activity1 创建时调用的方法
    @Override
    protected void onCreate(Bundle savedInstanceState) {
        super.onCreate(savedInstanceState);
        setContentView(R.layout.activity_main);
        Log.i("Activity1", "onCreate()");
        Button button = (Button)findViewById(R.id.button);
        button.setOnClickListener(new View.OnClickListener() {
            @Override
            public void onClick(View view) {
                Intent intent = new Intent(MainActivity.this,SecondActivity.class);
                startActivity(intent);
            }
        });
    }
}
```

以上就是单击事件的处理过程，后两种匿名内部类与主类中实现 OnClickListener 接口的方式在平常的 Android 开发中使用得较为普遍，所以需要熟练掌握。

## 3.2 Intent 的使用

### 3.2.1 Intent 浅析

在 Android 系统中，组件之间的通信需要使用到 Intent。Intent 中文翻译为"意图"，Intent 最常用的是绑定应用程序组件，并在应用程序之间进行通信。它一般用于启动 Activity、服务、发送广播等，承担了 Android 应用程序三大核心组件之间的通信功能。

使用 Intent 开启 Activity 时，对应的方法为 startActivity(Intent intent)和 startActivityForResult(Intent intent)；开启 Service 时，常用的有 ComponentName startService(Intent intent)和 boolean bindService(Intent service, ServiceConnection conn, int flags)。开启 BroadcastReceiver 方法有多种，就不一一列举了。

Android 中使用 Intent 的方式有两种，分别为显式 Intent 和隐式 Intent，接下来将在 3.2.2 节和 3.2.3 节详细介绍这两种方式。

### 3.2.2 显式 Intent

显式 Intent 就是在通过 Intent 启动 Activity 时，需要明确指定激活组件的名称，例如通过一个

Activity 启动另外一个 Activity 时,就可以通过这种方式。具体代码如下所示:

```
//创建 Intent 对象,指定启动的类名 SecondActivity
Intent intent = new Intent(MainActivity.this,SecondActivity.class);
//启动 Activity
startActivity(intent);
```

通过上述代码可以看出,使用显式 Intent 时,首先需要通过 Intent 的构造方法来创建 Intent 对象。构造方法有两个参数,分别为启动 Activity 的上下文和需要启动的 Activity 类名。除了通过指定类名的方式启动组件外,显式 Intent 还可以根据目标组件的包名、全路径来指定开启的组件。具体代码如下所示:

```
//setClassName("包名","类的全路径名称");
intent.setClassName("com.jxust.cn", "com.jxust.cn.chapter_shengtime");
//启动 Activity
startActivity(intent);
```

Activity 类提供了 startActivity(Intent intent)方法,该方法专门用于启动 Activity,它接收一个 Intent 参数,然后通过将构建好的 Intent 参数传入 startActivity()方法中来启动 Activity。

使用上述两种方式启动 Activity,能够在程序中很清晰地看到,其"意图"很明显,因此称为显式 Intent。

### 3.2.3 隐式 Intent

在程序中没有明确指定需要启动的 Activity,Android 系统会根据在 AndroidManifest.xml 文件当中设置的动作(Action)、类别(Category)、数据(Uri 和数据类型)来启动合适的组件。具体代码如下所示:

视频讲解

```
< activity android:name = ".MainActivity">
  < intent - filter >
<!-- 设置 action 属性,根据 name 设置的值来指定启动的组件 -->
    < action android:name = "android.intent.action.MAIN" />
    < category android:name = "android.intent.category.LAUNCHER" />
        </intent - filter >
    </activity >
```

在上述代码中,< action >标签指定了当前 Activity 可以响应的动作为 android.intent.action.MAIN,而< category >标签则包含了一些类别信息,只有当这两者中的内容同时匹配时,Activity 才会启动。使用隐式 Intent 启动 Activity 的具体代码如下:

```
Intent intent = new Intent();
intent.setAction("android.intent.action.MAIN");
StartActivity(intent);
```

通过以上的学习,已经初步了解了显式 Intent 和隐式 Intent 的使用。显式 Intent 启动组件时必须要指定组件的名称，一般只在本应用程序切换组件时使用。而隐式 Intent 使用的范围更广，不仅可以启动本应用程序内的组件，还可以开启其他应用的组件，如打开系统的照相机、图库等。

## 3.3 Activity 中的数据传递方式

在 Android 开发中，经常需要在 Activity 中进行数据传递，这里就需要使用 3.2 节讲到的 Intent 来实现 Activity 之间数据的传递。

使用 Intent 进行数据传递时只需要调用 putExtra()方法把数据存储进去即可。这个方法有两个参数，是一种"键值对"的形式，第一个参数为 key，第二个参数为 value。实现传递参数的具体代码如下：

```
//定义字符串变量存储一个值
String str = "android";
Intent intent = new Intent(this, SecondActivity.class);
//传递参数
intent.putExtra("receive_str", str);
startActivity(intent);
```

上述代码中将一个字符串变量 str 传递到 SecondActivity 中，然后需要在 SecondActivity 中接收这个参数。具体代码如下所示：

```
Intent intent = this.getIntent();
String receive_str = intent.getStringExtra("receive_st");
```

上面就是通过 Intent 传递和接收参数的一种简单方式，如果需要传递的参数比较多，就需要使用 putExtras()方法传递数据，该方法传递的是 Bundle 对象。具体代码如下所示：

```
Intent intent = new Intent(this, SecondActivity.class);
Bundle bundle = new Bundle();
bundle.putString("phone", "123456");
bundle.putString("sex", "男");
bundle.putString("age", "18");
intent.putExtras(bundle);
startActivity(intent);
```

上述代码使用 Bundle 对象传递参数，在 SecondActivity 中取出这些参数的具体代码如下所示：

```
Intent intent = this.getIntent();
Bundle bundle = intent.getExtras();
String phone = bundle.getString("phone");
```

在上述代码中，在接收 Bundle 对象封装的数据时，需要先接收对应的 Bundle 对象，然后再根据 key 取出 value。接下来将在 3.4 节讲解如何使用在布局文件中定义的各个组件以及如何进行数据的传递并显示。

## 3.4 用户注册案例功能实现

本节的用户注册布局与第 2 章的用户注册案例布局是一样的,本节主要讲的是如何使用布局中定义的组件以及数据传递与接收。

(1) Activity1 布局代码与第 2 章布局代码一样。

(2) Activity1 代码如下所示:

```java
public class MainActivity extends Activity implements View.OnClickListener,
RadioGroup.OnCheckedChangeListener{
    //定义字符串用来保存各个信息
    private String phone_str = "";
    private String paswd_str = "";
    //默认为选中"男性"
    private String sex_str = "男性";
    private String hobby_str = "1";
    private String city_str = "";
    //组件定义
    EditText phone_edit, paswd_edit;
    RadioGroup sex_group;
    RadioButton nan_but, nv_but;
    CheckBox play, read, music;
    Button register;
    Spinner spinner;
    @Override
    protected void onCreate(Bundle savedInstanceState) {
        super.onCreate(savedInstanceState);
        setContentView(R.layout.activity_main);
        //组件初始化
        phone_edit = (EditText)findViewById(R.id.phone);
        paswd_edit = (EditText)findViewById(R.id.paswd);
        sex_group = (RadioGroup)findViewById(R.id.sex);
        //添加监听事件
        sex_group.setOnCheckedChangeListener(this);
        nan_but = (RadioButton)findViewById(R.id.nan);
        read = (CheckBox)findViewById(R.id.read_book);
        play = (CheckBox)findViewById(R.id.play_ball);
        music = (CheckBox)findViewById(R.id.music);
        register = (Button)findViewById(R.id.register);
        //添加监听事件
        register.setOnClickListener(this);
        spinner = (Spinner)findViewById(R.id.spinner);
        final String[] city = new String[]{"北京", "上海", "武汉", "南京", "南昌", "信阳"};
        ArrayAdapter<String> adapter = newArrayAdapter<String>(this,
            android.R.layout.simple_list_item_1, city);
        spinner.setAdapter(adapter);
        //在"城市"下拉列表添加监听事件
        spinner.setOnItemSelectedListener(new Spinner.OnItemSelectedListener() {
            @Override
```

```
                public void onItemSelected(AdapterView<?> adapterView, View view, int i, long l) {
                    city_str = city[i];
                }
                @Override
                public void onNothingSelected(AdapterView<?> adapterView) {
                }
            });
        }
        @Override
        public void onClick(View view) {
            switch (view.getId()){
                case R.id.register:
                    //获取手机号和密码
                    phone_str = phone_edit.getText().toString();
                    paswd_str = paswd_edit.getText().toString();
                    //获取兴趣爱好即复选框的值
                    hobby_str = "";                    //清除上一次已经选中的选项
                    if(read.isChecked()){
                        hobby_str += read.getText().toString();
                    }if (play.isChecked()){
                        hobby_str += play.getText().toString();
                    }if(music.isChecked()){
                        hobby_str += music.getText().toString();
                    }
                    Intent intent = new Intent(this,SecondActivity.class);
                    Bundle bundle = new Bundle();
                    bundle.putString("phone", phone_str);
                    bundle.putString("paswd", paswd_str);
                    bundle.putString("sex", sex_str);
                    bundle.putString("hobby", hobby_str);
                    bundle.putString("city", city_str);
                    intent.putExtras(bundle);
                    startActivity(intent);
                    break;
            }
        }
        @Override
        public void onCheckedChanged(RadioGroup radioGroup, @IdRes int i) {
            //根据用户选择来改变 sex_str 的值
            sex_str = i == R.id.nan? "男性":"女性";
        }
    }
```

上述代码中主要是对 main_activity_layout.xml 中的组件进行初始化以及添加监听事件,然后对选择的数据进行传递。

(3) SecondActivity 负责接收数据并显示出来。在布局文件 second_lay.xml 中定义一个 TextView 负责显示数据。代码如下所示:

```
<?xml version = "1.0" encoding = "utf-8"?>
<LinearLayout xmlns:android = "http://schemas.android.com/apk/res/android"
```

```
        android:layout_width = "match_parent"
        android:layout_height = "match_parent">
    < TextView
        android:id = "@ + id/show_content"
        android:layout_width = "wrap_content"
        android:layout_height = "wrap_content"
        android:layout_weight = "1"
        android:text = "TextView" />
</LinearLayout >
```

(4) SecondActivity 的代码如下所示:

```
public class SecondActivity extends Activity {
    protected void onCreate(Bundle savedInstanceState) {
        super.onCreate(savedInstanceState);
        setContentView(R.layout.second_layout);
        Intent intent = this.getIntent();
        Bundle bundle = intent.getExtras();
        String phone = bundle.getString("phone");
        String paswd = bundle.getString("paswd");
        String sex = bundle.getString("sex");
        String hobby = bundle.getString("hobby");
        String city = bundle.getString("city");
        TextView show_text = (TextView)findViewById(R.id.show_content);
        show_text.setText("手机号为:" + phone + "\n" + "密码为:" + paswd + "\n" + "性别是:" +
            sex + "\n" + "爱好是:" + hobby + "\n" + "城市是:" + city);
    }
}
```

(5) 在 AndroidManifest.xml 文件中注册 SecondActivity,代码如下所示:

```
< activity android:name = ".SecondActivity"></activity >
```

上述代码编写完成以后,接下来输入手机号、密码,选择性别、爱好、城市,然后单击"注册"按钮,跳转到接收数据的界面。注册界面和接收数据界面分别如图 3-10 和图 3-11 所示。

图 3-10　注册界面

图 3-11　接收数据界面

以上就是用户注册的详细介绍，这里包含了组件的使用、数据的传递和接收、按钮的单击事件等，这些都是进行 Android 开发的基本知识，需要熟练掌握和应用。

## 3.5　本章小结

本章首先讲解 Activity 的基本知识，包含从 Activity 的概念、生命周期，到后面的从一个 Activity 跳转到另外一个 Activity 生命周期中方法的执行过程；然后讲解 Intent 的使用以及数据的传递和接收；最后结合一个用户注册的实例讲解组件的使用以及监听事件的处理，这些都需要开发者熟练掌握。

## 3.6　课后习题

1. 简述一个 Activity 跳转到另一个 Activity 时，两个 Activity 生命周期方法的执行过程。
2. 编写一个程序，要求在第一个界面中输入两个数字，在第二个界面中显示第一个界面中输入的两个数字的和、差、积、商。

# 第 4 章

# Android 事件处理

**学习目标**
- 掌握 Android 基于监听的事件处理。
- 掌握 Android 基于回调的事件处理。
- 掌握 AnsyncTask 异步类的功能与用法。

当进行 Android 程序开发时,首先对用户界面进行编程,然后就是用户与界面的交互了。当用户对界面进行各种操作时,程序需要为用户提供响应的动作,这种响应的动作就需要通过事件处理来完成。

Android 为用户提供了两种方式的事件处理:基于回调的事件处理与基于监听的事件处理。本章将会详细介绍 Android 的事件处理方式。

## 4.1 Android 事件处理机制

UI 编程通常都会伴随事件处理,Android 也不例外,它提供了两种方式的事件处理:基于回调的事件处理和基于监听的事件处理。

对于基于监听的事件处理而言,主要就是为 Android 界面组件绑定特定的事件监听器;对于基于回调的事件处理而言,主要做法是重写 Android 组件特定的回调函数,Android 大部分界面组件都提供了事件响应的回调函数,主要是重写这些回调函数。

## 4.2 基于监听的事件处理

基于监听的事件处理相比于基于回调的事件处理,是更具"面向对象"性质的事件处理方式。在监听器模型中,主要涉及 3 类对象。

- 事件源(Event Source):产生事件的来源,通常是各种组件,如按钮等。

- 事件(Event)：封装了界面组件上发生的特定事件的具体信息，如果监听器需要获取界面组件上所发生事件的相关信息，一般通过事件 Event 对象来传递。
- 事件监听器(Event Listener)：负责监听事件源发生的事件，并对不同的事件做相应的处理。

基于监听的事件处理机制是一种委派式(Delegation)事件处理方式：事件源将整个事件处理委托给事件监听器；当该事件源发生指定的事件时，就通知委托的事件监听器，由事件监听器来处理它。

下面将通过一个单击按钮显示用户在 EditText 输入内容的例子来讲述基于监听的事件处理方式。

首先创建一个新的 Android 项目 Chapter4_Listener，界面布局代码如下所示：

```xml
<?xml version = "1.0" encoding = "utf-8"?>
<LinearLayout xmlns:android = "http://schemas.android.com/apk/res/android"
    xmlns:app = "http://schemas.android.com/apk/res-auto"
    xmlns:tools = "http://schemas.android.com/tools"
    android:layout_width = "match_parent"
    android:layout_height = "match_parent"
    android:orientation = "vertical"
    tools:context = "com.jxust.cn.chapter4_listener.MainActivity">
    <EditText
        android:id = "@+id/edittext1"
        android:layout_width = "match_parent"
        android:layout_height = "50dp"
        android:hint = "请输入内容:"
        android:textColor = "@android:color/black"
        android:textSize = "18sp"/>
    <Button
        android:id = "@+id/bt1"
        android:layout_width = "match_parent"
        android:layout_height = "50dp"
        android:textColor = "@android:color/black"
        android:text = "获取 EditText 内容"/>
</LinearLayout>
```

上面程序中定义的按钮将会作为事件源，接下来程序将会为该按钮绑定一个事件监听器。

MainActivity 类主要是获取应用程序的按钮，为其添加监听事件并且处理，具体代码如下所示：

```java
public class MainActivity extends Activity
implements View.OnClickListener {
    EditText editText;
    @Override
    protected void onCreate(Bundle savedInstanceState) {
        super.onCreate(savedInstanceState);
        setContentView(R.layout.activity_main);
        editText = (EditText)findViewById(R.id.edittext1);
```

```
        Button bt1 = (Button)findViewById(R.id.bt1);
        bt1.setOnClickListener(this);            //为按钮绑定事件监听器
    }
    //实现监听器类必须实现的方法,该方法将会作为事件处理器
    @Override
    public void onClick(View view) {
        String str = editText.getText().toString();
        Toast.makeText(this,str,Toast.LENGTH_SHORT).show();
    }
}
```

上面的程序通过 MainActivity 实现 OnClickListener 接口,然后重写该方法,该方法作为事件处理器来处理按钮的单击事件。当界面中的按钮被单击时,出现 EditText 中输入的内容。运行效果如图 4-1 所示。

从上面的程序中可以看出,基于监听的事件处理机制的步骤如下:

(1) 获取普通界面的组件即事件源。

(2) 实现事件的监听器类,该监听器类是一个特殊的 Java 类,必须实现一个 XxxListener 接口。

(3) 调用事件源的 setXxxListener()方法将事件监听器对象注册给事件源。

图 4-1　基于监听的事件处理效果图

当事件源上发生指定事件时,Android 会触发事件监听器,由事件监听器调用相应的方法来处理事件。

关于什么样的类可以作为监听器类,在 3.1.4 节已经讲解过,例如内部类作为事件监听器类、外部类作为事件监听器类、匿名内部类作为事件监听器类、Activity 作为事件监听器类等。

## 4.3　基于回调的事件处理

视频讲解

相比基于监听的事件处理模型,基于回调的事件处理模型要简单些,在该模型中,事件源和事件监听器是合一的,也就是说,没有独立的事件监听器存在。当用户在 GUI 组件上触发某事件时,由该组件自身特定的函数负责处理该事件。通常通过重写 Override 组件类的事件处理函数实现事件的处理。

为了使用回调机制类处理 GUI 组件上所发生的事件,需要通过继承 GUI 组件类,并重写该类的事件处理方法来实现。

为了实现回调机制的事件处理,Android 为所有的 GUI 组件都提供了一些事件处理的回调方法,例如对 View 来说,该类包含如下方法:

- boolean onKeyDown(int keycode,KeyEvent event)——用户在该组件上按下某个按键时触发的方法。
- boolean onKeyLongPress(int keycode,KeyEvent event)——用户在该组件上长按某个组件

时触发的方法。
- boolean onKeyUp(int keycode,KeyEvent event)——用户在该组件上松开某个按键时触发的方法。

下面将通过一个自定义按钮的实现类来讲解基于回调的事件处理机制。

首先自定义实现类的代码如下所示：

```java
public class TestButton extends Button {
    public TestButton(Context context, AttributeSet attrs) {
        super(context, attrs);
    }
    /* 重写 onTouchEvent()触碰事件的回调方法 */
    @Override
    public boolean onTouchEvent(MotionEvent event) {
        Log.i("测试 CallBack", "我是 Button, 你触碰了我： " + event.getAction());
        Toast.makeText(getContext(),"我是 MyButton, 你触碰了我：" +
            event.getAction(), Toast.LENGTH_SHORT).show();
        return true; //返回 true,表示事件不会向外层(即父容器)扩散
    }
}
```

在上述代码中，通过自定义一个 TestButton 类继承 Button 类，然后重写该类的 onTouchEvent()方法来负责处理屏幕上按钮的触摸事件。

布局文件中使用了这个自定义 View，具体代码如下所示：

```xml
<?xml version = "1.0" encoding = "utf-8"?>
<LinearLayout xmlns:android = "http://schemas.android.com/apk/res/android"
    tools:context = "com.jxust.cn.chapter4_callback.MainActivity">
    <com.jxust.cn.chapter4_callback.TestButton
        android:layout_width = "match_parent"
        android:layout_height = "wrap_content"
        android:textSize = "18sp"
        android:text = "测试基于回调的事件处理机制"/>
</LinearLayout>
```

上述代码中加粗部分的代码使用 MyButton 组件，接下来 Java 程序也不需要再为该按钮绑定事件监听器，因为该按钮自己重写了 onTouchEvent()方法，这意味着该按钮将会自己处理相应的事件。

上面的代码运行效果如图 4-2 和图 4-3 所示。

通过上面的学习，发现对于基于监听的事件处理机制来说，事件源和事件监听器是分离的，当事件源上发生特定事件时，该事件交给事件监听器负责处理；对于基于回调的事件处理机制来说，事件源和事件监听器是统一的，当事件源发生特定事件时，该事件还是由事件源自己负责处理。

图 4-2　单击按钮

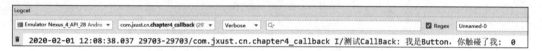

图 4-3　基于回调的事件处理效果图

## 4.4　AnsyncTask 异步类的功能与用法

视频讲解

　　Android 的 UI 线程主要负责处理用户的按键事件、触屏事件等。因此其他阻塞 UI 线程的操作不应该在主线程中进行。

　　为了避免 UI 线程失去响应的问题，Android 程序采用将耗时操作放在新线程中完成的方式，但是新线程可能需要动态更新 UI 组件，比如获取网络资源操作放在新线程中完成。但由于新线程不允许直接更新 UI 组件，为了解决这个问题，Android 提供了以下几种方式：

- 使用 Hanlder 实现线程之间的通信。
- View.post(Runnable)。
- Activity.runOnUiThread(Runnable)。

　　上述方式使用起来有点复杂，采用异步任务（AsyncTask）则可以进一步简化操作。相对来说，异步任务 AsyncTask 更轻量级一些，适用于简单的异步任务，例如获取网络数据、动态更改 UI 界面等。

　　AsyncTask< Params, Progress, Result >是一个抽象类，通常用于被继承，继承时需要指定如下 3 个泛型参数：

- Params——启动任务执行的输入参数的类型。
- Progress——后台任务完成的进度值的类型。

- Result——后台任务执行完成以后返回结果的类型。

使用 AsyncTask 的步骤如下：

（1）创建 AsyncTask 的子类，并指定参数类型。如果某个参数不需要，则指定为 Void 类型。

（2）实现 AsyncTask 的方法，如 doInBackground(Params…)，后台线程将要完成的功能，一般用于获取网络资源等耗时性的操作；又如 onPostExecute(Result result)，在 doInBackground() 方法执行完以后，系统会自动调用 onPostExecute() 方法，并接收其返回值。这里一般负责更新 UI 线程等操作。

（3）调用 AsyncTask 子类的实例的 execute(Params… params) 方法，执行耗时操作。

接下来通过一个从网络下载图片的具体例子来讲解如何使用 AsyncTask 类。

首先是界面布局，这里采用相对布局的方式，只需要一个 ImageView 显示图片和一个 ProgressBar 以查看是否下载完成，代码如下所示：

```xml
<RelativeLayout xmlns:android = "http://schemas.android.com/apk/res/android"
    xmlns:tools = "http://schemas.android.com/tools"
    android:id = "@+id/RelativeLayout1"
    android:layout_width = "match_parent"
    android:layout_height = "match_parent"
    tools:context = "com.jxust.cn.chapter4_asynctask.MainActivity" >
    <ImageView
        android:id = "@+id/imageView1"
        android:layout_width = "fill_parent"
        android:layout_height = "fill_parent"
        android:layout_alignParentLeft = "true"
        android:layout_alignParentTop = "true"
        />
    <ProgressBar
        android:id = "@+id/progressBar1"
        android:visibility = "gone"
        style = "?android:attr/progressBarStyleLarge"
        android:layout_width = "wrap_content"
        android:layout_height = "wrap_content"
        android:layout_centerHorizontal = "true"
        android:layout_centerVertical = "true" />
</RelativeLayout>
```

上述代码中使用了 ProgressBar，并且使用了 android:visibility="gone" 设置了隐藏属性。

MainActivity 里的代码如下所示：

```java
public class MainActivity extends AppCompatActivity {
    private ImageView mImageView = null;
    private ProgressBar mProgressBar = null;
    private String URLs = "https://ss0.bdstatic.com/5aV1bjqh_Q23odCf/static/
        superman/img/logo/bd_logo1_31bdc765.png";
    @Override
    protected void onCreate(Bundle savedInstanceState) {
```

```java
        super.onCreate(savedInstanceState);
        setContentView(R.layout.activity_main);
        //实例化组件
        this.mImageView = (ImageView) findViewById(R.id.imageView1);
        this.mProgressBar = (ProgressBar) findViewById(R.id.progressBar1);
        //实例化异步任务
        ImageDownloadTask task = new ImageDownloadTask();
        //执行异步任务
        task.execute(URLs);
    }
    class ImageDownloadTask extends AsyncTask<String,Void,Bitmap> {
        @Override
        protected Bitmap doInBackground(String... params) {
            Bitmap bitmap = null;                       //待返回的结果
            String url = params[0];                     //获取 URL
            URLConnection connection;                   //网络连接对象
            InputStream is;                             //数据输入流
            try {
                connection = new URL(url).openConnection();
                is = connection.getInputStream();       //获取输入流
                BufferedInputStream buf = new BufferedInputStream(is);
                //解析输入流
                bitmap = BitmapFactory.decodeStream(buf);
                is.close();
                buf.close();
            } catch (MalformedURLException e) {
                e.printStackTrace();
            } catch (IOException e) {
                e.printStackTrace();
            }
            //返回给后面调用的方法
            return bitmap;
        }
        @Override
        protected void onPreExecute() {
            //显示等待圆环
            mProgressBar.setVisibility(View.VISIBLE);
        }
        @Override
        protected void onPostExecute(Bitmap result) {
            //下载完毕,隐藏等待圆环
            mProgressBar.setVisibility(View.GONE);
            mImageView.setImageBitmap(result);
        }
    }
}
```

最后是要在 AndroidManifest、XML 里加上网络访问权限,代码如下所示:

```
<uses-permission android:name="android.permission.INTERNET"></uses-permission>
```

Google 公司为保证用户数据和设备的安全，6.0 以上版本要求默认使用加密连接，URL 的链接采用 HTTPS 请求。关于网络访问权限在第 9 章会详细介绍。上述代码运行结果如图 4-4 所示。

MainActivity 中有一个内部类：ImageDownloadTask，这个内部类用来下载指定 URL 的图片，并把图片在 ImageView 里显示出来。

将内部类异步处理部分（即 doInBackground()方法）看作一个异步图片下载器，传入这个下载器的是图片的 URL，下载器传出的是图片，下载过程中不需要知道图片加载进度，所以 3 个泛型参数的类型分别为 String、Void 和 Bitmap。

当启动异步任务时，先执行 onPreExecute()方法，所以可以在这个方法里显示 progressBar；然后启动子线程执行 doInBackground()，并将参数传给此方法，这个方法里会进行网络操作，下载图片并将其转为 Bitmap 返回，返回后子线程也结束了；最后执行的是 onPostExecute()方法，这个方法获取的参数是异步处理后的结果，即下载好的图片，ImageView 显示下载好的图片，并隐藏 ProgressBar 等待圆环。

图 4-4　异步下载网络图片

视频讲解

## 4.5 【项目案例 3】"橙一色"App 注册验证

本章前几节介绍了 Android 的事件处理机制，本节结合这些知识，实现"橙一色"App 的注册功能。即当用户进入注册界面，输入相关信息后，单击"注册"按钮实现信息的保存。注册过程要求实现对密码与确认密码是否一致、地址是否选择、账号或者密码是否填写等进行验证，验证通过后，跳转到个人信息展示的界面。注册验证的功能自定义了一个 validateRegister()方法，写在了 RegisterActivity 类中。具体代码如下所示：

```
public class RegisterActivity extends AppCompatActivity implements
    View.OnClickListener, RadioGroup.OnCheckedChangeListener {
    private EditText usernameEdit, passwordEdit, surePasswordEdit;
    private TextView cityText;
    private CityPicker cityPicker;
    private Button regButton;
    private RadioGroup sexGroup;
    private String sexStr = "男";
    @Override
    protected void onCreate(@Nullable Bundle savedInstanceState) {
        super.onCreate(savedInstanceState);
        setContentView(R.layout.user_register);
```

```java
        init();
    }
    //界面组件初始化
    private void init() {
        cityText = findViewById(R.id.reg_province);
        cityText.setOnClickListener(this);
        usernameEdit = findViewById(R.id.reg_username);
        passwordEdit = findViewById(R.id.reg_password);
        surePasswordEdit = findViewById(R.id.reg_sure_password);
        regButton = findViewById(R.id.reg_register);
        regButton.setOnClickListener(this);
        sexGroup = findViewById(R.id.sex);
        sexGroup.setOnCheckedChangeListener(this);
    }
    @Override
    public void onClick(View v) {
        switch (v.getId()) {
            case R.id.reg_province:
                initCityPicker();
                cityPicker.show();
                break;
            case R.id.reg_register:
                //注册验证方法
                validateRegister();
                break;
        }
    }
    //初始化城市选择器
    public void initCityPicker() {
        cityPicker = new CityPicker.Builder(RegisterActivity.this)
                .textSize(16)
                .title("地址选择")
                .backgroundPop(0xa0000000)
                .titleBackgroundColor("#EFB81C")
                .titleTextColor("#000000")
                .backgroundPop(0xa0000000)
                .confirTextColor("#000000")
                .cancelTextColor("#000000")
                .province("江西省")
                .city("赣州市")
                .district("章贡区")
                .textColor(Color.parseColor("#000000"))
                .provinceCyclic(true)
                .cityCyclic(false)
                .districtCyclic(false)
                .visibleItemsCount(7)
                .itemPadding(10)
                .onlyShowProvinceAndCity(false)
```

```java
            .build();
    cityPicker.setOnCityItemClickListener(new CityPicker.OnCityItemClickListener() {
        @Override
        public void onSelected(String... strings) {
            String province = strings[0];
            String city = strings[1];
            String district = strings[2];
            cityText.setText(String.format("%s %s %s", province, city, district));
        }
        @Override
        public void onCancel() {
        }
    });
}
//注册验证
public void validateRegister() {
    Intent intent = new Intent(RegisterActivity.this, UserActivity.class);
    String username = usernameEdit.getText().toString();
    String password = passwordEdit.getText().toString();
    String surePassword = surePasswordEdit.getText().toString();
    String city = cityText.getText().toString();
    //判断两次密码是否输入一致
    if (password.equals(surePassword)) {
        //这里也可以再进行其他的验证,如是否符合指定格式要求等,符合就进行插入数据库操作
        if (!username.equals("") || !password.equals("")) {
            if (!city.equals("")) {
                Bundle bundle = new Bundle();
                bundle.putString("username", username);
                bundle.putString("password", password);
                bundle.putString("sex", sexStr);
                bundle.putString("city", city);
                intent.putExtras(bundle);
                startActivity(intent);
            } else {
                    Toast.makeText(RegisterActivity.this,"请选择地址",
                        Toast.LENGTH_SHORT).show();
            }
        } else {
            Toast.makeText(RegisterActivity.this,"账号或密码未填写",
                Toast.LENGTH_SHORT).show();
        }
    } else {
        Toast.makeText(RegisterActivity.this, "两次密码输入不一致",
            Toast.LENGTH_SHORT).show();
    }
}
@Override
public void onCheckedChanged(RadioGroup group, int checkedId) {
```

```
        //根据用户选择来改变性别的值
        sexStr = checkedId == R.id.reg_man ? "男" : "女";
    }
}
```

上述代码中加粗的validateRegister()方法部分,对用户名和密码、性别的选择、城市的选择是否合法进行了验证,当都验证通过时,即跳转到个人信息界面。假设注册一个叫作jsust的用户,选择性别及地区等,然后单击"注册"按钮,运行结果分别如图4-5和图4-6所示。

图4-5　注册界面

图4-6　个人信息界面

个人信息界面主要包含头像、卡券、信息展示、通用和设置等内容,头像为一个ImageView居中显示,下面的每一行由一个TextView和"右箭头"的ImageView组成,TextView里包含一个图像和一个文字信息。这里只列出布局的部分代码,具体的可以查看本章节对应案例的源代码。

```xml
<?xml version = "1.0" encoding = "utf-8"?>
< LinearLayout xmlns:android = "http://schemas.android.com/apk/res/android"
    android:layout_width = "match_parent"
    android:layout_height = "match_parent"
    android:background = "#E8E8E8"
    android:orientation = "vertical">
    <!-- 头像 -->
    < ImageView
        android:id = "@ + id/user_icon"
        android:layout_width = "wrap_content"
```

```xml
        android:layout_height = "wrap_content"
        android:layout_marginTop = "20dp"
        android:layout_gravity = "center_horizontal"
        android:src = "@drawable/user_icon" />
...
    <!-- 用户名 -->
    <LinearLayout
        android:id = "@+id/user_username_line"
        android:layout_width = "match_parent"
        android:layout_height = "wrap_content"
        android:background = "#FFF"
        android:layout_marginTop = "30dp"
        android:orientation = "horizontal">
        <TextView
            android:id = "@+id/user_username"
            android:layout_height = "35dp"
            android:layout_weight = "1"
            android:drawableLeft = "@drawable/user_account"
            android:textColor = "#000"
            android:text = "用户名:"
            android:gravity = "center_vertical"
            android:drawablePadding = "8dp"
            android:textSize = "14sp"
            android:layout_marginLeft = "5dp"
            android:layout_width = "0dp" />
        <ImageView
            android:layout_width = "25dp"
            android:layout_height = "25dp"
            android:layout_gravity = "center"
            android:src = "@drawable/arrow_right" />
    <!-- 卡券、性别等布局类似于用户名布局(此处省略) -->
    <Button
        android:layout_width = "fill_parent"
        android:layout_height = "40dp"
        android:layout_marginLeft = "10dp"
        android:layout_marginRight = "10dp"
        android:textSize = "18sp"
        android:layout_marginTop = "150dp"
        android:text = "退出登录"
        android:textColor = "#FFFFFF"
        android:layout_gravity = "center"
        android:background = "@drawable/button_login"/>
</LinearLayout>
```

## 4.6 本章小结

本章主要讲解 Android 程序的两种事件处理机制,因为当开发一个应用界面时,用户需要与界面进行各种交互,界面需要通过事件处理来对用户的操作提供响应动作。接着讲解 Android 开发时使用较多的 AsyncTask(异步任务),这就解决了获取网络图片等耗时操作问题,避免了 UI 线程阻塞等。最后介绍了项目案例"橙一色"App 中注册验证的功能实现。本章的内容较为重要,需要熟练掌握。

## 4.7 课后习题

1. 说明 Android 两种事件处理机制的不同。
2. 对于 Android 的两种事件处理机制,分别写一个案例测试,了解其执行过程。
3. 编写一个 Android 程序,使用 AsyncTask 实现获取网页 HTML 代码,并且使用 TextView 显示。

# 第 5 章

# Fragment 基础

**学习目标**
- 掌握 Fragment 的生命周期。
- 掌握 Fragment 的创建。
- 掌握 Fragment 与 Activity 之间的通信。

随着移动设备的快速发展,平板电脑也越来越普及,而平板电脑与手机的最大差别就在于屏幕的大小。为了同时兼顾手机和平板电脑的开发,自 Android 3.0(API level 11)开始引入了 Fragment。接下来将对 Fragment 进行详细的介绍。

## 5.1 Fragment 概述

Fragment 翻译为中文就是"碎片"的意思,它是一种嵌入到 Activity 中使用的 UI 片段。一个 Activity 里面可以包含一个或多个 Fragment,而且一个 Activity 可以同时展示多个 Fragment。使用它能够让程序能够更加合理地利用拥有大屏幕空间的移动设备,因此 Fragment 在平板电脑上应用非常广泛。

Fragment 与 Activity 类似,也拥有自己的布局与生命周期,但是它的生命周期会受到它所在的 Activity 的生命周期的控制。例如,当 Activity 暂停时,它所包含的 Fragment 也会暂停;当 Activity 被销毁时,该 Activity 内的 Fragment 也会被销毁;当该 Activity 处于活动状态时,开发者才可独立地操作 Fragment。

为了更加清楚地讲解 Fragment 的功能,接下来将会通过一个图例来说明,如图 5-1 所示。

从图 5-1 可以看出,在一般的手机上或者平板电脑竖屏情况下,Fragment1 需要嵌入到 Activity1 中,Fragment2 需要嵌入到 Activity2 中;如果在平板电脑横屏的情况下,则可以把两个 Fragment 同时嵌入到 Activity1 中,这样的布局既节约了空间,也会更美观。

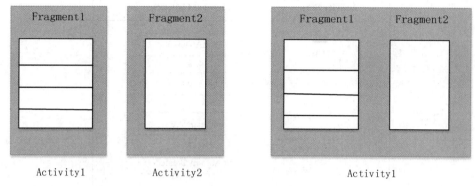

图 5-1　Fragment 的功能

## 5.2　Fragment 的生命周期

通过第 3 章的学习，我们知道 Activity 生命周期有 3 种状态，分别是运行状态、暂停状态和停止状态。Fragment 与 Activity 非常相似，其生命周期也会经历这几种状态。接下来将会详细介绍这几种状态。

运行状态：当嵌入该 Fragment 的 Activity 处于运行状态时，并且该 Fragment 是可见的，那么该 Fragment 是处于运行状态的。

暂停状态：当嵌入该 Fragment 的 Activity 是处于暂停状态时，那么该 Fragment 也是处于暂停状态的。

停止状态：当嵌入该 Fragment 的 Activity 处于停止状态时，那么该 Fragment 也会进入停止状态。或者通过调用 FragmentTransation 的 reMove()、rePlace()方法将 Fragment 从 Activity 中移除。

Fragment 必须是依存于 Activity 而存在的，因此 Activity 的生命周期会直接影响到 Fragment 的生命周期。图 5-2 很好地说明了两者生命周期的关系。

可以看到，Fragment 比 Activity 多了几个额外的生命周期回调方法。

- onAttach(Activity)：当 Fragment 与 Activity 发生关联时调用。
- onCreateView(LayoutInflater, ViewGroup, Bundle)：创建该 Fragment 的视图（加载布局）时调用。
- onActivityCreated(Bundle)：当 Activity（与 Fragment 相关联）的 onCreate()方法返回时调用。
- onDestroyView()：与 onCreateView()相对应，当与该 Fragment 关联的视图被移除时调用。
- onDetach()：与 onAttach()相对应，当 Fragment 与 Activity 关联被取消时调用。

以上就是 Fragment 的生命周期与 Activity 的生命周期之间的关系，接下来将会讲解如何创建 Fragment 以及 Fragment 之间的通信。

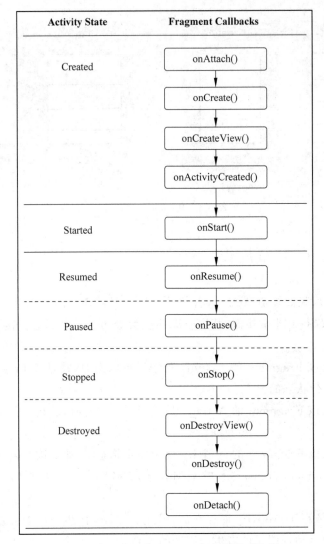

图 5-2　Fragment 和 Activity 生命周期对比图

视频讲解

## 5.3　Fragment 的创建

　　Fragment 的创建与 Activity 的创建类似,要创建一个 Fragment 必须要创建一个类继承自 Fragment。Android 系统提供了两个 Fragment 类,分别是 android.app.Fragment 和 android.support.v4.app.Fragment。继承前者只能兼容 Android 4.0 以上的系统,继承后者可以兼容更低的版本。Android 8.0 版本中 androidx.fragment.app.Fragment 取代了 android.app.Fragment,因而新建的 Fragment 默认继承 androidx.fragment.app.Fragment。接下来讲解 Fragment 的创建过程。

　　(1) 新建一个左侧的碎片布局文件 fragment_left.xml,代码如下:

```
<?xml version = "1.0" encoding = "utf-8"?>
< LinearLayout xmlns:android = "http://schemas.android.com/apk/res/android"
```

```
    xmlns:tools = "http://schemas.android.com/tools"
    android:layout_width = "match_parent"
    android:layout_height = "match_parent"
    android:orientation = "vertical"
    tools:context = ".LeftFragment">
    <Button
        android:id = "@+id/button"
        android:layout_width = "wrap_content"
        android:layout_height = "wrap_content"
        android:layout_gravity = "center_horizontal"
        android:text = "单击我"
        />
</LinearLayout>
```

(2) 新建一个右侧的碎片布局文件 fragment_right.xml,代码如下:

```
<?xml version = "1.0" encoding = "utf-8"?>
<LinearLayout xmlns:android = "http://schemas.android.com/apk/res/android"
    xmlns:tools = "http://schemas.android.com/tools"
    android:layout_width = "match_parent"
    android:layout_height = "match_parent"
    android:background = "@android:color/darker_gray"
    android:orientation = "vertical"
    tools:context = ".RightFragment">
    <ImageView
        android:layout_width = "80dp"
        android:layout_height = "80dp"
        android:layout_marginTop = "10dp"
        android:layout_marginLeft = "60dp"
        android:src = "@mipmap/ic_launcher"/>
</LinearLayout>
```

(3) 接着新建一个 LeftFragment 类,继承自 Fragment,代码如下:

```
public class Left_Fragment extends Fragment {
    @Override
    public View onCreateView(LayoutInflater inflater, ViewGroup container, Bundle savedInstanceState) {
        View view = inflater.inflate(R.layout.fragment_left, container, false);
        return view;
    }
}
```

(4) 这里仅仅是重写了 Fragment 的 onCreateView()方法,然后这个方法中通过 LayoutInflater 的 inflate()方法将刚才定义的 fragment_left 布局动态加载进来,然后再新建一个 SecondRightFragment 类,继承自 Fragment,代码如下:

```java
public class SecondRightFragment extends Fragment {
    @Override
    public View onCreateView(LayoutInflater inflater, ViewGroup container, Bundle savedInstanceState) {
        View view = inflater.inflate(R.layout.fragment_second_right, container, false);
        return view;
    }
}
```

(5) 新建 fragment_second_right.xml 文件,用来显示单击按钮时更换的界面,代码如下所示:

```xml
<LinearLayout xmlns:android = "http://schemas.android.com/apk/res/android"
    android:layout_width = "match_parent"
    android:layout_height = "match_parent"
    android:background = "@android:color/holo_blue_dark"
    android:orientation = "vertical" >
    <Button
        android:layout_marginTop = "10dp"
        android:layout_width = "match_parent"
        android:layout_height = "wrap_content"
        android:text = "我是左边单击出来的哦" />
</LinearLayout>
```

(6) 新建 SecondRightFragment 作为另一个右侧碎片,代码如下所示:

```java
public class SecondRightFragment extends Fragment {
    @Override
    public View onCreateView(LayoutInflater inflater, ViewGroup container, Bundle savedInstanceState) {
        View view = inflater.inflate(R.layout.fragment_second_right, container, false);
        return view;
    }
}
```

(7) 修改 activity_main.xml,代码如下所示:

```xml
<?xml version = "1.0" encoding = "utf-8"?>
<LinearLayout xmlns:android = "http://schemas.android.com/apk/res/android"
    xmlns:app = "http://schemas.android.com/apk/res-auto"
    xmlns:tools = "http://schemas.android.com/tools"
    android:layout_width = "match_parent"
    android:layout_height = "match_parent"
    android:orientation = "horizontal"
    tools:context = ".MainActivity">
    <fragment
        android:id = "@+id/left_fragment"
        android:name = "com.jxust.cn.chapter5_fragment_create.LeftFragment"
        android:layout_width = "0dp"
        android:layout_height = "match_parent"
```

```xml
            android:layout_weight = "1"
            tools:layout = "@layout/fragment_left" />
    <FrameLayout
        android:id = "@+id/right_layout"
        android:layout_width = "0dp"
        android:layout_height = "match_parent"
        android:layout_weight = "1" >
        <!-- 可以在这个容器中动态加载 Fragment -->
        <fragment
            android:id = "@+id/right_fragment"
            android:name = "com.jxust.cn.chapter5_fragment_create.RightFragment"
            android:layout_width = "match_parent"
            android:layout_height = "match_parent"
            tools:layout = "@layout/fragment_right" />
    </FrameLayout>
</LinearLayout>
```

（8）可以看到，现在将右侧碎片放在了一个 FrameLayout 中，这是 Android 中最简单的一种布局，它没有任何的定位方式，所有的组件都会摆放在布局的左上角。由于这里仅需要在布局里放入一个碎片，因此非常适合使用 FrameLayout。之后将在代码中替换 FrameLayout 里的内容，从而实现动态添加碎片的功能。修改 MainActivity 中的代码，如下所示：

```java
public class MainActivity extends AppCompatActivity implements View.OnClickListener {
    Button button;
    @Override
    protected void onCreate(Bundle savedInstanceState) {
        super.onCreate(savedInstanceState);
        setContentView(R.layout.activity_main);
        button = (Button)findViewById(R.id.button);
        button.setOnClickListener(this);
    }
    @Override
    public void onClick(View view) {
        switch (view.getId()){
          case R.id.button:
                SecondRightFragment secFragment = new SecondRightFragment();
                FragmentManager fragmentManager = getSupportFragmentManager();
                FragmentTransaction transaction = fragmentManager.beginTransaction();
                transaction.replace(R.id.right_layout, secFragment);
                transaction.commit();
                break;
          default:
                break;
        }
    }
}
```

可以看到,首先给左侧碎片中的按钮注册了一个单击事件,然后将动态添加碎片的逻辑都放在了单击事件中进行。结合代码可以看出,动态添加碎片主要分为如下 5 步:

(1) 创建待添加的碎片实例。
(2) 获取到 FragmentManager,在活动中可以直接调用 getFragmentManager()方法得到。
(3) 开启一个事务,通过调用 beginTransaction()方法开启。
(4) 向容器内加入碎片,一般使用 replace()方法实现,需要传入容器的 ID 和待添加的碎片实例。
(5) 提交事务,调用 commit()方法来完成。

这样就完成了在活动中动态添加碎片的功能,运行程序,可以看到启动界面如图 5-3 所示,然后单击一下按钮,效果如图 5-4 所示。

图 5-3　启动界面

图 5-4　单击按钮结果图

上述代码成功实现了向活动中动态添加碎片的功能,不过这时按下键盘上的返回键程序就会直接退出。如果这里想模仿类似返回栈的效果,可以通过 FragmentTransaction 中提供的一个 addToBackStack()方法,将一个事务添加到返回栈中,修改 MainActivity 中的代码,如下所示:

```java
@Override
    public void onClick(View view) {
        switch (view.getId()){
            case R.id.button:
                SecondRightFragment secFragment = new SecondRightFragment();
                FragmentManager fragmentManager = getSupportFragmentManager();
                FragmentTransaction transaction = fragmentManager.beginTransaction();
                transaction.replace(R.id.right_layout, secFragment);
                transaction.addToBackStack(null);
                transaction.commit();
                break;
            default:
                break;
        }
    }
```

这里在事务提交之前调用了 FragmentTransaction 的 addToBackStack()方法,它可以接收一个名字用于描述返回栈的状态,一般传入 null 即可。现在重新运行程序,并单击按钮将 SecondRightFragment 添加到活动中,然后按下返回键,会发现程序并没有退出,而是回到了 RightFragment 界面,再次按下返回键程序才会退出。

## 5.4 Fragment 与 Activity 之间的通信

视频讲解

由于 Fragment 与 Activity 都是各自存在于一个独立的类中,它们之间并没有明显的方式进行直接通信。在实际的开发过程当中,经常需要在 Activity 中获取 Fragment 实例或者在 Fragment 中获取 Activity 实例。接下来将详细地讲解 Fragment 和 Activity 之间的通信。

- 在 Activity 中获取 Fragment 实例。

为了实现 Fragment 和 Activity 之间的通信,FragmentManager 提供了一个 findFragmentById() 的方法,专门用于从布局文件中获取 Fragment 实例。该方法有一个参数,它代表 Fragment 在 Activity 布局中的 ID。例如在布局文件中指定 SecondFragment 的 ID 为 R. id. second_fragmnet,这时就可以使用 getFragmentManager(). findRagmentById(R. id. second_fragmnet)方法得到 SecondFragment 的实例。

为了更好地理解,下面通过一段代码讲解,具体代码如下所示:

```
SecondFragment second_frag = (SecondFragment) getFragmentManager()
                    .findFragmentById(R.id.second_fragment);
```

以上就是在 Activity 中获取 Fragment 实例的代码。

- 在 Fragment 中获取 Activity 实例。

在 Fragment 中获取 Activity 实例对象,可以通过在 Fragment 中调用 getActivity()方法来获取与当前 Fragment 相关联的 Activity 实例对象。例如在 MainActivity 中添加了 SecondFragment,那么就可以通过在 Fragment 中调用 getActivity()来获取 MainActivity 实例对象。具体代码如下所示:

```
MainActivity main = (MainActivity)getActivity();
```

获取到 Activity 中的实例以后,就可以调用该 Activity 中的方法了。当 Fragment 需要使用 Context 对象时,也可以使用该方法。

以上就是在 Activity 中获取 Fragment 实例和在 Fragment 中获取 Activity 实例对象的具体代码。接下来将通过具体的例子来讲解两者之间的通信方式。

为了更好地掌握 Fragment 与 Activity 之间的通信,接下来介绍一个左边显示新闻标题、右边展示单击新闻标题以后出现新闻的具体内容的例子,具体的操作步骤如下所示。

### 1. 创建新闻展示项目

首先创建新闻展示项目,然后修改 activity_main. xml 中的布局代码,因为需要展示标题和对应的内容,所以需要添加两个 FrameLayout,后边将会被 Fragment 所代替。具体如下所示:

```xml
<?xml version = "1.0" encoding = "utf-8"?>
<LinearLayout xmlns:android = "http://schemas.android.com/apk/res/android"
    xmlns:app = "http://schemas.android.com/apk/res-auto"
    xmlns:tools = "http://schemas.android.com/tools"
    android:layout_width = "match_parent"
    android:layout_height = "match_parent"
    tools:context = "com.jxust.cn.chapter5_news.MainActivity"
    android:orientation = "horizontal">
    <!-- 标题 -->
    <FrameLayout
        android:id = "@+id/settitle"
        android:layout_width = "0dp"
        android:layout_weight = "1"
        android:layout_height = "match_parent">
    </FrameLayout>
    <!-- 内容 -->
    <FrameLayout
        android:id = "@+id/setcontent"
        android:layout_width = "0dp"
        android:layout_weight = "2"
        android:layout_height = "match_parent">
    </FrameLayout>
</LinearLayout>
```

### 2. 创建两个 Fragment 布局文件

由于需要实现在一个 Activity 中展示两个 Fragment,因此需要创建相应的 Fragment 的布局。用来展示新闻标题的布局文件 title_layout.xml 代码如下:

```xml
<?xml version = "1.0" encoding = "utf-8"?>
<LinearLayout xmlns:android = "http://schemas.android.com/apk/res/android"
    android:layout_width = "match_parent"
    android:layout_height = "match_parent">
    <!-- 用来展示新闻标题列表 -->
    <ListView
        android:id = "@+id/titlelist"
        android:layout_width = "match_parent"
        android:layout_height = "wrap_content">
    </ListView>
</LinearLayout>
```

用来展示右边标题和内容的布局文件 content_layout.xml 代码如下:

```xml
<?xml version = "1.0" encoding = "utf-8"?>
<LinearLayout xmlns:android = "http://schemas.android.com/apk/res/android"
    android:layout_width = "match_parent"
    android:layout_height = "match_parent"
```

```xml
        android:orientation = "vertical">
    <TextView
        android:id = "@ + id/show_title"
        android:layout_width = "match_parent"
        android:layout_height = "wrap_content"
        android:textSize = "20sp"
        android:text = "显示新闻标题" />
    <TextView
        android:id = "@ + id/show_content"
        android:layout_width = "match_parent"
        android:layout_marginTop = "20dp"
        android:layout_height = "wrap_content"
        android:textSize = "16sp"
        android:text = "显示新闻内容" />
</LinearLayout>
```

### 3. 创建 ListView 中每一项的内容布局

由于左边的新闻标题采用了 ListView，因此需要创建一个显示 ListView 中每一项的布局文件，title_item_layout.xml 文件的代码如下所示：

```xml
<?xml version = "1.0" encoding = "utf-8"?>
<LinearLayout xmlns:android = "http://schemas.android.com/apk/res/android"
    android:layout_width = "match_parent"
    android:layout_height = "match_parent">
    <!-- 用来展示新闻标题列表 -->
    <TextView
        android:id = "@ + id/titles"
        android:layout_width = "wrap_content"
        android:layout_height = "wrap_content"
        android:textSize = "16sp"/>
</LinearLayout>
```

### 4. 创建显示标题的 Fragment 类文件

创建一个 SetTitleFragment 类文件继承自 Fragment 类，用来显示左边的新闻标题，具体代码如下所示：

```java
import android.support.v4.app.Fragment;
import android.view.LayoutInflater;
import android.view.View;
import android.view.ViewGroup;
import android.widget.AdapterView;
import android.widget.BaseAdapter;
import android.widget.ListView;
import android.widget.TextView;
public class SetTitleFragment extends Fragment {
    private View view;
```

```java
private String[] title;
private String[][] contents;
private ListView listView;
public View onCreateView(LayoutInflater inflater, final ViewGroup container, Bundle
                    savedInstanceState){
    view = inflater.inflate(R.layout.title_layout, container, false);
    //获取 Activity 实例对象
    MainActivity activity = (MainActivity)getActivity();
    //获取 Activity 中的标题
    title = activity.getTitle();
    //获取 Activity 中的标题和内容
    contents = activity.getSettingText();
    if (view!= null){
        init();
    }
    //为 ListView 添加监听
    listView.setOnItemClickListener(new AdapterView.OnItemClickListener() {
        @Override
        public void onItemClick(AdapterView<?> adapterView, View view, int i, long l) {
            //通过 Activity 实例获取另一个 Fragment 对象
            setContentFragment content =
            (setContentFragment)((MainActivity)getActivity()).
            getSupportFragmentManager().findFragmentById(R.id.setcontent);
            content.setText(contents[i]);
        }
    });
    return view;
}
private void init() {
    listView = (ListView)view.findViewById(R.id.titlelist);
    if (title!= null){
        listView.setAdapter(new MyAdapter());
    }
}
//适配器
class MyAdapter extends BaseAdapter{
    @Override
    public int getCount() {
        return title.length;
    }
    @Override
    public Object getItem(int i) {
        return title[i];
    }
    @Override
    public long getItemId(int i) {
        return i;
    }
```

```java
        @Override
        public View getView(int i, View view, ViewGroup viewGroup) {
            view = View.inflate(getActivity(), R.layout.title_item_layout, null);
            TextView titletext = (TextView)view.findViewById(R.id.titles);
            titletext.setText(title[i]);
            return view;
        }
    }
}
```

**5. 创建显示标题和内容的 Fragment 类文件**

创建一个类 SetContentFragment 继承自 Fragment 类，然后编写相应的逻辑代码，用来显示左边单击以后出现的内容，具体代码如下所示：

```java
package com.jxust.cn.chapter5_news;
import android.app.Activity;
import android.os.Bundle;
import android.support.v4.app.Fragment;
import android.view.LayoutInflater;
import android.view.View;
import android.view.ViewGroup;
import android.widget.TextView;
public class SetContentFragment extends Fragment {
    private View view;
    private TextView text1, text2;
    public void onAttach(Activity activity){
        super.onAttach(activity);
    }
    public View onCreateView(LayoutInflater inflater, ViewGroup container, Bundle savedInstanceState){
        //获取布局文件
        view = inflater.inflate(R.layout.content_layout, container, false);
        if (view!= null){
            init();
        }
        //获取 Activity 中设置的文字
        setText(((MainActivity)getActivity()).getSettingText()[0]);
        return view;
    }
    private void init() {
        text1 = (TextView)view.findViewById(R.id.show_title);
        text2 = (TextView)view.findViewById(R.id.show_content);
    }
    public void setText(String[] text) {
        text1.setText(text[0]);
        text2.setText(text[1]);
    }
}
```

### 6. 编写 MainActivity 中代码

编写好两个 Fragment 类的代码以后，就需要在 MainActivity 中添加，具体代码如下所示：

```java
import android.support.v4.app.FragmentActivity;
import android.os.Bundle;
public class MainActivity extends FragmentActivity {
    //设置标题
    private String title[] = {"标题一","标题二","标题三"};
    private String settingText [][] = {{"标题一","标题一的内容"},{"标题二","标题二的内容"},
                                       {"标题三","标题三的内容"}};
    //获取标题数组的方法
    public String[] getTilte(){
        return tilte;
    }
    //获取标题和内容
    public String[][] getSettingText(){
        return settingText;
    }
    @Override
    protected void onCreate(Bundle savedInstanceState) {
        super.onCreate(savedInstanceState);
        setContentView(R.layout.activity_main);
        //创建 Fragment
        setTitleFragment TitleFragment = new setTitleFragment();
        setContentFragment ContentFragment = new setContentFragment();
        //获取事物
        android.support.v4.app.FragmentManager fragmentManager
            = getSupportFragmentManager();
        android.support.v4.app.FragmentTransaction transaction
            = fragmentManager.beginTransaction();
        //添加 Fragment
        transaction.replace(R.id.settitle,TitleFragment);
        transaction.replace(R.id.setcontent,ContentFragment);
        //提交事物
        transaction.commit();
    }
}
```

图 5-5　Fragment 与 Activity 通信案例图

### 7. 测试运行

以上就是 Activity 与 Fragment 之间的通信过程。上述代码实现了如图 5-5 所示的界面。

从图 5-5 可以看出，当单击屏幕左侧的标题以后，右侧的界面也会跟着显示对应的标题和内容，这就说明了本实例实现了 Activity 与 Fragment 之间的通信以及 Fragment 与 Fragment 之间的通信。这部分内容需要开发者熟练掌握。

## 5.5 【项目案例 4】 "橙一色"App 商品展示功能实现

视频讲解

"橙一色"App 是一款水果在线销售的 App,在前面的项目案例中已经完成了登录界面设计和注册验证功能展示,本节将实现商品展示的功能,根据选择的分类来展示对应的水果信息。利用到的知识是本章学习的 Fragment。

(1) 添加商品查询布局。本节不对布局进行大的改变,在第 12 章的 App 功能优化中将会对登录或者注册成功以后进入的界面进行变动,实现更多的功能。商品查询的操作设计目前就在个人信息布局上添加一个线性布局,用来设计单击事件。具体内容如下:

```xml
<!-- 查看商品 -->
<LinearLayout
    android:id = "@ + id/user_searchProduct"
    android:layout_width = "match_parent"
    android:layout_height = "wrap_content"
    android:background = "#FFF"
    android:layout_marginTop = "20dp"
    android:orientation = "horizontal">
    <TextView
        android:layout_height = "35dp"
        android:layout_weight = "1"
        android:drawableLeft = "@drawable/search"
        android:textColor = "#000"
        android:text = "查看商品"
        android:gravity = "center_vertical"
        android:drawablePadding = "8dp"
        android:textSize = "14sp"
        android:layout_marginLeft = "5dp"
        android:layout_width = "0dp" />
    <ImageView
        android:layout_width = "25dp"
        android:layout_height = "25dp"
        android:layout_gravity = "center"
        android:src = "@drawable/arrow_right" />
</LinearLayout>
```

(2) 为线性布局添加 ID 属性,在 Activity 中为其添加单击事件,用来跳转到查询商品的界面。商品查询界面的设计主要包含了标题、分类的列表和商品信息的展示。content_category.xml 具体的内容如下:

```xml
<?xml version = "1.0" encoding = "utf-8"?>
<LinearLayout xmlns:android = "http://schemas.android.com/apk/res/android"
    android:layout_width = "match_parent"
    android:layout_height = "match_parent"
    android:background = "#E8E8E8"
```

```xml
    android:orientation = "vertical">
    <!-- 标题 -->
    <LinearLayout
        android:layout_width = "match_parent"
        android:layout_height = "50dp"
        android:background = "#EFB81C"
        android:orientation = "horizontal">
        <ImageView
            android:id = "@+id/category_return"
            android:layout_width = "wrap_content"
            android:layout_height = "wrap_content"
            android:layout_gravity = "center"
            android:src = "@drawable/arrow_left" />
        <TextView
            android:layout_width = "0dp"
            android:layout_height = "wrap_content"
            android:layout_marginTop = "12dp"
            android:layout_weight = "1"
            android:gravity = "center"
            android:text = "商品种类"
            android:textColor = "#FFF"
            android:textSize = "20sp" />
    </LinearLayout>
    <!-- 分类标题和内容 -->
    <LinearLayout
        android:layout_width = "match_parent"
        android:layout_height = "match_parent"
        android:orientation = "horizontal">
        <!-- 标题 -->
        <ListView
            android:id = "@+id/category_title_list"
            android:layout_width = "0dp"
            android:layout_weight = "1"
            android:layout_height = "match_parent" />
        <!-- 内容 -->
        <FrameLayout
            android:id = "@+id/category_detail"
            android:layout_width = "0dp"
            android:layout_height = "match_parent"
            android:layout_weight = "3" />
    </LinearLayout>
</LinearLayout>
```

(3) 上述布局代码中,主要包含 ListView 和 Fragment,分别用来展示分类信息和对应的水果信息。ListView 当中,自定义了一个分类标题 category_title.xml 文件,具体内容如下:

```xml
<?xml version = "1.0" encoding = "utf-8"?>
<LinearLayout xmlns:android = "http://schemas.android.com/apk/res/android"
```

```xml
    android:layout_width = "match_parent"
    android:layout_height = "match_parent"
    android:orientation = "vertical">
    <TextView
        android:id = "@ + id/categor_titles"
        android:layout_width = "match_parent"
        android:gravity = "center"
        android:layout_height = "wrap_content"
        android:textSize = "18sp"
        android:text = "标题" />
</LinearLayout>
```

（4）水果信息主要包括水果的图标、名字以及价格。定义一个布局文件用于展示水果信息。category_detail_content.xml 内容如下：

```xml
<?xml version = "1.0" encoding = "utf-8"?>
<LinearLayout xmlns:android = "http://schemas.android.com/apk/res/android"
    android:layout_width = "match_parent"
    android:layout_height = "match_parent"
    android:orientation = "vertical">
    <ImageView
        android:id = "@ + id/category_product_image"
        android:layout_width = "90dp"
        android:layout_height = "90dp"
        android:layout_gravity = "center" />
    <TextView
        android:id = "@ + id/category_product_name"
        android:layout_width = "wrap_content"
        android:layout_height = "30dp"
        android:layout_gravity = "center"
        android:layout_marginTop = "2dp"
        android:textColor = "#050505"
        android:textSize = "16sp" />
    <TextView
        android:id = "@ + id/category_product_price"
        android:layout_width = "wrap_content"
        android:layout_height = "30dp"
        android:layout_gravity = "center"
        android:layout_marginTop = "2dp"
        android:gravity = "center"
        android:textColor = "#050505"
        android:textSize = "16sp" />
</LinearLayout>
```

（5）上述代码主要完成了布局的设计，接下来需要定义一个 Fragment 和一个 Activity 用来操作商品的选择，定义一个 CategoryActivity 用来初始化界面组件以及添加监听事件。具体代码如下：

```java
public class CategoryActivity extends Activity {
    public OnChangeListener onchangedListener;
    private List<Product> productList;
    private List<String> productCategory = new ArrayList<>();
    private ListView titleList;
    private MyAdapter adapter;

    @Override
    public void onCreate(@Nullable Bundle savedInstanceState) {
        super.onCreate(savedInstanceState);
        setContentView(R.layout.content_category);
        initData();
        init();
        SetDetailFragment fragment = new SetDetailFragment();
        FragmentManager fragmentManager = getFragmentManager();
        FragmentTransaction transaction = fragmentManager.beginTransaction();
        transaction.replace(R.id.category_detail, fragment);
        transaction.commit();
        titleList.setOnItemClickListener((parent, view, position, id) -> {
            adapter.setSelectedPosition(position);
            adapter.notifyDataSetInvalidated();
            if (onchangedListener != null) {
                onchangedListener.changeText(productList.get(position));
            }
        });
    }
    public void setOnChangeListener(OnChangeListener onChangeListener) {
        this.onchangedListener = onChangeListener;
    }
    public interface OnChangeListener {
        void changeText(Product product);
    }
```

```xml
<!-- 初始化数据 -->
```

```java
    private void initData() {
        productList = new ArrayList<>();
        productCategory.add("橘子");
        productCategory.add("橙子");
        productCategory.add("柚子");
        Product product = new Product();
        product.setImageUrlId(R.drawable.juzi);
        product.setProductName("橘子");
        product.setProductPrice(new BigDecimal("9.9"));
        Product product1 = new Product();
        product1.setImageUrlId(R.drawable.chengzi);
        product1.setProductName("橙子");
        product1.setProductPrice(new BigDecimal("29.9"));
        Product product2 = new Product();
        product2.setImageUrlId(R.drawable.youzi);
```

```java
        product2.setProductName("柚子");
        product2.setProductPrice(new BigDecimal("19.9"));
        productList.add(product);
        productList.add(product1);
        productList.add(product2);
    }
    <!--初始化组件-->
    private void init() {
        titleList = findViewById(R.id.category_title_list);
        adapter = new MyAdapter(productCategory, CategoryActivity.this);
        titleList.setAdapter(adapter);
    }
```

（6）从上述代码可以看到，首先初始化数据，然后为显示标题的 ListView 设置适配器，并为其添加元素的单击事件。创建 MyAdapter 适配器的代码如下所示：

```java
public class MyAdapter extends BaseAdapter {
    private List<String> productCategory;
    private LayoutInflater layoutInflater;
    private int selectionPosition = -1;
    public MyAdapter(List<String> productCategory, Context context) {
        this.productCategory = productCategory;
        this.layoutInflater = LayoutInflater.from(context);
    }
    @Override
    public int getCount() {
        return productCategory.size();
    }
    @Override
    public Object getItem(int position) {
        return productCategory.get(position);
    }
    @Override
    public long getItemId(int position) {
        return position;
    }
    @Override
    public View getView(int position, View convertView, ViewGroup parent) {
        ViewHolder viewHolder = null;
        if (convertView == null) {
            viewHolder = new ViewHolder();
            convertView = layoutInflater.inflate(R.layout.category_list_item, null);
            Log.i("adapts", "getView: " + convertView);
            viewHolder.tv = convertView.findViewById(R.id.categor_titles);
            convertView.setTag(viewHolder);
        } else {
            viewHolder = (ViewHolder) convertView.getTag();
```

```java
        }
        viewHolder.tv.setText(productCategory.get(position));
        if (selectionPosition == position) {
            viewHolder.tv.setBackgroundColor(Color.YELLOW);
        } else {
            viewHolder.tv.setBackgroundColor(Color.WHITE);
        }
        return convertView;
    }
    public void setSelectedPosition(int position) {
        this.selectionPosition = position;
    }
    class ViewHolder {
        TextView tv;
    }
}
```

适配器主要包含了标题的列表，默认选择元素的坐标、布局文件。自定义一个类继承自 BaseAdapter，然后重写其内部的方法。最主要的是 getView() 方法的重写，首先获取自定义的布局文件，然后根据传过来的值对其对应的组件进行赋值操作。

(7) 自定义一个类用来继承 Fragment，然后重写 onCreateView() 方法来为布局文件内的组件进行赋值。定义 SetDetailFragment 的内容如下所示：

```java
public class SetDetailFragment extends Fragment {
    private View view;
    private ImageView imageView;
    private TextView nameText, priceText;
    @SuppressLint("SetTextI18n")
    @Nullable
    @Override
    public View onCreateView(@NonNull LayoutInflater inflater, @Nullable ViewGroup container,
                @Nullable Bundle savedInstanceState) {
        view = inflater.inflate(R.layout.categoty_detail_content, container, false);
        if (view != null) {
            init();
        }
        CategoryActivity categoryActivity = (CategoryActivity) getActivity();
        Objects.requireNonNull(categoryActivity).setOnChangeListener(product -> {
            Log.i("sss", "onCreateView: " + product.getProductName());
            imageView.setBackgroundResource(product.getImageUrlId());
            nameText.setText(product.getProductName());
            priceText.setText(product.getProductPrice().toString());
        });
        return view;
    }
    <!-- 内容组件初始化 -->
```

```
private void init() {
    imageView = view.findViewById(R.id.category_product_image);
    nameText = view.findViewById(R.id.category_product_name);
    priceText = view.findViewById(R.id.category_product_price);
}
}
```

至此,关于水果分类选择的功能已基本实现,后续会对其进行更好的调整。本节商品展示的实现效果如图 5-6 所示。

图 5-6　水果查询界面

## 5.6　本章小结

本章主要讲解了 Fragment 的概念、生命周期、与 Activity 之间的通信方式以及 Fragment 和 Fragment 之间的通信方式,这些知识在平板电脑开发或者考虑到屏幕兼容性开发中经常使用,需要开发者熟练掌握并应用到实际的项目中。

## 5.7　课后习题

1. 说明 Fragment 的生命周期。
2. Android 提供了 Navigation Drawer Activity 和 Tabbed Activity,可以实现与 5.4 节新闻展示项目同样的功能,试编程实现。

# Android 数据存储

**学习目标**
- 了解数据存储方式的特点。
- 掌握文件存储、XML 文件、SharedPreferences 的使用。
- 掌握 SQLite 数据库的使用。
- 掌握 JSON 类型的数据使用。

在 Android 开发中,大多数应用程序都需要存储一些数据,例如用户信息的保存、商品信息的展示、图片的存储等。Android 中数据存储方式有文件存储、SharedPreferences、SQLite 数据库、网络存储、ContentProvider 5 种。当前 Android 端的数据和界面一般是分离的,数据的交互通常是通过 JSON 的格式,所以需要重点掌握。本章将对文件存储、XML 的序列化与解析、SharedPreferences、SQLite 数据库以及 JSON 类型的数据结构进行详细的讲解。

## 6.1 数据存储方式简介

Android 中的 5 种数据存储方式有不同的特点,接下来将详细的介绍这 5 种数据存储方式的特点。

文件存储:把要存储的文件,如音乐、图片等以 I/O 流的形式存储在手机内存或者 SD 卡中。

SharedPreferences:和 XML 文件存储的类型相似,都是以键值对的形式存储数据,常用这种方式存储用户登录时的用户名和密码等信息。

SQLite:一个软件库,实现了自给自足的、无服务器的、零配置的、事务性的 SQL 数据库引擎。SQLite 是在世界上最广泛部署的 SQL 数据库引擎。它是一个轻量级、跨平台的数据库,通常用于存储用户的信息等。

网络存储:把数据存储到服务器中,使用的时候可以连接网络,然后从网络上获取信息,这样保证了信息的安全性等。

ContentProvider:内容提供者,是 Android 中的四大组件之一。主要用于对外共享数据,也就是通过ContentProvider把应用中的数据共享给其他应用访问,其他应用可以通过ContentProvider对指定应用中的数据进行操作。ContentProvider分为系统的和自定义的,系统的ContentProvider也就是如联系人、相册等数据。

以上就是Android的5种数据存储方式简介,需要注意的是,如果需要把数据共享给其他的应用程序使用,可以使用文件存储、SharedPreferences、ContentProvider,一般使用ContentProvider较好。

## 6.2 文件存储

Android中的文件存储与Java中的文件存储类似,都是以I/O流的形式把数据存储到文件中;而不同点在于Android中的文件存储分为外部存储和内部存储两种,接下来分别介绍两种存储方式。

### 6.2.1 外部存储

视频讲解

外部存储就是指把文件存储到一些外部设备上,例如SD卡、设备内的存储卡等,属于永久性存储方式。使用这种类型存储的文件可以共享给其他的应用程序使用,也可以被删除、修改、查看等,它不是一种安全的存储方式。

由于外部存储方式一般存放在外部设备中,所以在使用之前要先检查外围设备是否存在。在Android中使用Environment.getExternalStorageState()方法来查看外部设备是否存在,当外部设备存在时,就可以使用FileInputStream、FileOutputStream、FileReader、FileWriter对象来读写外部设备中的文件。

向外部设备存储文件的具体代码如下所示:

```java
//检查外部设备是否存在
    String environment = Environment.getExternalStorageState();
    if (Environment.MEDIA_MOUNTED.equals(environment)) {
        //外部设备可以进行读写操作
        File sd_path = Environment.getExternalStorageDirectory();
        File file = new File(sd_path,"test.txt");
        String str = "Android";
        FileOutputStream fi_out;
        try{
            fi_out = new FileOutputStream(file);
            fi_out.write(str.getBytes());
            fi_out.close();
        }
        catch(Exception exception){
            exception.printStackTrace();
        }
    }
```

上述代码实现了向SD卡中的test.txt文件中存储一个字符串信息。首先使用Environment

.getExternalStorageState()方法来检查是否存在外部设备,然后使用 Environment.getExternalStorageDirectory()获取 SD 卡的路径,由于不同的手机上 SD 卡的路径可能不同,使用这种方式可避免把路径写死而出现找不到路径的错误。

Android 为了保证应用程序的安全性,无论是读取还是写入操作,都需要添加权限,否则程序将会出错。在 AndroidManifest.xml 文件中添加如下所示的权限代码:

```
< uses - permission android:name = "android.permission.WRITE_EXTERNAL_STORAGE" />
< uses - permission android:name = "android.permission.READ_EXTERNAL_STORAGE" />
```

由于 Android 6.0 以后设置了权限管理机制,如果 Android Studio 建立的模拟器版本是 6.0 以上的,则需要用户手动授权。打开模拟器的"设置"(setting),单击 Apps & notifications,找到当前的项目 Chapter6_SDCard,选择 Permissions,打开 App 的存储权限,具体操作如图 6-1 所示。

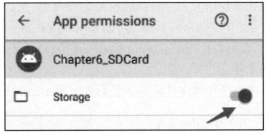

图 6-1　设置读写 SD 卡的权限

完成上述设置后,程序就可以操作 SD 卡中的数据了。写入的数据保存在 Device File Explore 的 sdcard 路径下,如图 6-2 所示。

图 6-2　SD 卡的文件

## 6.2.2 内部存储保存用户注册信息

内部存储是指将应用程序的数据以文件的形式存储在应用程序的目录下(data/data/< packagename/files/目录下>),这个文件属于该应用程序私有,如果其他应用程序想要操作本应用程序的文件,就需要设置权限。内部存储的文件随着应用程序的卸载而删除,随着应用程序的生成而创建。

接下来通过一个使用文件存储用户注册信息的案例讲解内部存储的使用过程。

**1. 创建项目**

创建一个 Chapter6_File 的 Android 项目,创建完成以后,修改布局文件的内容,设计如图 6-3 所示的用户界面。

图 6-3 用户界面

实现上述界面的布局文件 activity_main 的代码如下所示:

```xml
<?xml version = "1.0" encoding = "utf-8"?>
< LinearLayout xmlns:android = "http://schemas.android.com/apk/res/android"
    xmlns:app = "http://schemas.android.com/apk/res-auto"
    xmlns:tools = "http://schemas.android.com/tools"
    android:layout_width = "match_parent"
    android:layout_height = "match_parent"
    tools:context = "com.jxust.cn.chapter6_file.MainActivity"
    android:orientation = "vertical">
    < EditText
        android:id = "@ + id/zhanghao"
        android:layout_width = "match_parent"
        android:layout_height = "wrap_content"
        android:hint = "请输入账号"/>
    < EditText
        android:id = "@ + id/paswd"
        android:layout_width = "match_parent"
        android:layout_height = "wrap_content"
        android:inputType = "numberPassword"
        android:hint = "请输入密码"/>
    < Button
        android:id = "@ + id/save_mes"
        android:layout_width = "match_parent"
        android:layout_height = "wrap_content"
        android:background = "#5CACEE"
        android:textColor = "#FFFFFF"
        android:textSize = "16sp"
        android:text = "保存用户信息"/>
    < Button
        android:id = "@ + id/show_mes"
        android:layout_width = "match_parent"
```

```
            android:layout_height = "wrap_content"
            android:layout_marginTop = "15dp"
            android:background = "#5CACEE"
            android:textColor = "#FFFFFF"
            android:textSize = "16sp"
            android:text = "读取用户信息"/>
</LinearLayout>
```

上述界面中定义了两个 EditText 用来输入用户的账号和密码,定义了两个按钮用来保存输入的信息和读取保存到文件当中的信息。

### 2. 编写 MainActivity 中的代码,实现保存和读取功能

上述界面设计完成以后,接下来就需要在 MainActivity 中初始化上述布局的组件,然后为按钮添加监听以实现用户信息的保存和读取,具体代码如下所示:

```java
public class MainActivity extends Activity implements View.OnClickListener {
    public EditText username, paswd;
    public Button save, show;
    @Override
    protected void onCreate(Bundle savedInstanceState) {
        super.onCreate(savedInstanceState);
        setContentView(R.layout.activity_main);
        init();
    }
    //组件初始化方法
    public void init(){
        username = (EditText)findViewById(R.id.zhanghao);
        paswd = (EditText)findViewById(R.id.paswd);
        save = (Button)findViewById(R.id.save_mes);
        show = (Button)findViewById(R.id.show_mes);
        //为按钮添加监听
        save.setOnClickListener(this);
        show.setOnClickListener(this);
    }
    @Override
    public void onClick(View view) {
        switch (view.getId()){
            case R.id.save_mes:
                //获取输入的账号和密码
                String user_str = username.getText().toString();
                String paswd_str = paswd.getText().toString();
                String user_mes = "用户名为:" + user_str + "\n" + "密码为" + paswd_str;
                FileOutputStream fi_out;
                try {
                    //保存输入的数据
                    fi_out = openFileOutput("user_mes.txt",MODE_PRIVATE);
                    fi_out.write(user_mes.getBytes());
```

```
                fi_out.close();
            }catch (Exception e){
                e.printStackTrace();
            }
        Toast.makeText(this,"用户信息已保存", Toast.LENGTH_SHORT).show();
            break;
        case R.id.show_mes:
            //存储读取到的信息
            String mes = "";
            try {
                FileInputStream fi_input;
                fi_input = openFileInput("user_mes.txt");
                byte[] buffer = new byte[fi_input.available()];
                fi_input.read(buffer);
                mes = new String(buffer);
                fi_input.close();
            }catch (Exception e){
                e.printStackTrace();
            }
            Toast.makeText(this, mes, Toast.LENGTH_SHORT).show();
            break;
        default:
            break;
        }
    }
}
```

上述代码中,openFileOutput(String name,int mode)方法用于打开输出流,将数据存储到文件中。参数 name 代表文件名,参数 mode 表示文件的操作权限。mode 有以下几种取值:

MODE_PRIVATE——默认的操作权限,只能被当前应用程序所读写。

MODE_APPEND——可以添加文件的内容。

MODE_WORLD_READABLE——可以被其他程序所读取,安全性较低。

MODE_WORLD_WRITEABLE——可以被其他的程序所写入,安全性低。

openFileInput(String name)方法用于获取文件输入流对象,然后将存储在缓冲区 buffer 的数据赋值给字符串 mes 变量。

上述代码把用户输入的账号和密码保存到 user_mes.txt 文件,并且实现了从文件当中读取并显示的功能,重点在于对按钮监听事件的处理。

### 3. 程序运行

完成了布局文件和 MainActivity 代码的编写以后,运行程序,单击"保存用户信息"按钮就会把用户信息保存到 user_mes.txt 文件中,并且提示用户保存成功;单击"读取用户信息"按钮就会从 user_mes.txt 文件中读取保存的信息,然后通过 Toast 的形式显示出来。具体的实现如图 6-4 和图 6-5 所示。

图 6-4 保存用户信息

图 6-5 读取用户信息

**4. 查看 user_mes.txt 文件是否存在**

为了确定上述代码是否产生了 user_mes.txt 文件,可以在 data/data/目录下查找。在 Android Studio 中查看文件的操作如下所示。

- 选择 View→Tool Windows→Device File Explorer 命令,默认在编写 Activity 窗体的右侧也有 Device File Explorer 快捷图标,打开 Device File Explorer,显示正在运行的模拟器,分别如图 6-6 和图 6-7 所示。

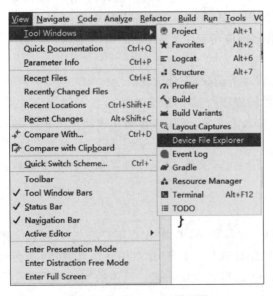

图 6-6 打开 Android 模拟器

图 6-7 打开 File Explorer

- 打开 Device File Explorer，然后查找文件，在/data/data/com.jxust.cn.chapter6_file/files/目录可以看到生成了一个 user_mes.txt 文件，这就说明上述代码成功实现了文件保存用户名的功能，如图 6-8 所示。

上面主要讲解了文件存储的两种存储方式，然后通过一个使用文件存储的方式保存用户信息的案例来具体讲解文件存储的使用，核心就是使用 I/O 流进行读写操作，这部分内容需要开发者熟练掌握。

图 6-8 查看文件

## 6.3 XML 文件的序列化与解析

在 Android 的开发中，XML 是非常常用的一种封装数据的形式，从服务器中获取数据也经常是这种形式的，所以学会生成和解析 XML 是非常有用的，接下就将对这两个知识点进行详细的介绍。

### 6.3.1 XML 序列化

XML 的序列化就是将对象类型的数据保存在 XML 文件中，又叫持久化。要将数据序列化，首先要创建与该 XML 相对应的 XML 文件生成器，然后将要存入的对象类型的数据转换为 XML 文件。XML 序列化的代码如下所示：

```
<!-- XML 文件生成器 -->
XmlSerializer serializer = Xml.newSerializer();
FileOutputStream fos = null;
<!-- 设置文件编码格式 -->
    serializer.setOutput(fos, "utf-8");
<!-- 开始文档,参数分别为字符编码和是否保持独立 -->
    serializer.startDocument("utf-8", true);
    serializer.startTag(null, "persons");              //开始标签
for (Person person: list) {
        serializer.startTag(null, "person");
        serializer.attribute(null, "id", person.getId() + "");
        serializer.startTag(null, "name");             //开始标签
        serializer.text(person.getName());             //文本内容
        serializer.endTag(null, "name");               //结束标签
serializer.endTag(null, "person");
}
serializer.endTag(null, "persons");                    //结束标签
serializer.endDocument();                              //结束 XML 文档
```

上述代码通过 XmlSerializer 对象可以设置 XML 文件的编码格式，然后向文件写入 XML 文件的开始标志"<?xml version＝"1.0" encoding＝"utf-8"?>"代码。然后通过设置开始节点、开始标签、添加内容、结束标签、结束节点完成 XML 文件的生成。下面通过一个保存用户信息的案例来讲解 XML 文件的序列化。

## 6.3.2 XML 序列化实例

6.3.1 节简单地介绍了 XML 文件序列化的步骤,接下来将通过一个具体的实例——保存用户登录的信息来讲解 XML 文件序列化的使用,具体的步骤如下。

**1. 修改 activity_main.xml 布局文件的代码**

修改 activity_main.xml 布局文件的代码,实现如图 6-9 所示的界面。

图 6-9  XML 文件序列化界面

activity_main.xml 的布局代码如下所示:

```xml
<?xml version = "1.0" encoding = "utf-8"?>
<LinearLayout xmlns:android = "http://schemas.android.com/apk/res/android"
    xmlns:app = "http://schemas.android.com/apk/res-auto"
    xmlns:tools = "http://schemas.android.com/tools"
    android:layout_width = "match_parent"
    android:layout_height = "match_parent"
    tools:context = "com.jxust.cn.chapter6_xml.MainActivity"
    android:orientation = "vertical">
    <EditText
        android:id = "@+id/username"
        android:layout_width = "match_parent"
        android:layout_height = "wrap_content"
        android:hint = "请输入用户名:"/>
    <EditText
        android:id = "@+id/paswd"
        android:layout_width = "match_parent"
        android:layout_height = "wrap_content"
android:inputType = "numberPassword"
        android:hint = "请输入密码:"/>
    <Button
        android:id = "@+id/button"
        android:layout_width = "match_parent"
        android:layout_height = "wrap_content"
        android:text = "XML 文件序列化" />
</LinearLayout>
```

## 2. 创建 Person 类

创建与 XML 文件对应的 Person 类,该类封装了用户的两个属性——用户名和密码,具体代码如下所示:

```java
public class Person {
    private String username;
    private String paswd;
    public Person(String username, String paswd) {
        this.username = username;
        this.paswd = paswd;
    }
    public void setUsername(String username) {
        this.username = username;
    }
    public String getUsername() {
        return username;
    }
    public void setPaswd(String paswd) {
        this.paswd = paswd;
    }
    public String getPaswd() {
        return paswd;
    }
    @Override
    public String toString() {
      return "Person{" + "username = '" + username + '\'' + ", paswd = '" + paswd + '\'' + '}';
    }
}
```

## 3. 编写 MainActivity 代码

编写 MainActivity 代码,为按钮添加监听事件,单击"XML 文件序列化"按钮以后,将数据保存到 SD 卡中的 person_mes.xml 文件中。具体代码如下所示:

```java
public class MainActivity extends AppCompatActivity implements View.OnClickListener {
    private EditText username,paswd;
    private Button button;
    @Override
    protected void onCreate(Bundle savedInstanceState) {
        super.onCreate(savedInstanceState);
        setContentView(R.layout.activity_main);
        init();
    }
    <!-- 组件初始化方法 -->
    public void init(){
        username = (EditText)findViewById(R.id.username);
        paswd = (EditText)findViewById(R.id.paswd);
```

```java
            button = (Button)findViewById(R.id.button);
            button.setOnClickListener(this);
    }
    @Override
    public void onClick(View view) {
        switch (view.getId()){
            case R.id.button:
                //XML文件序列化操作
                //取得输入的数据
                String name_str = username.getText().toString();
                String paswd_str = paswd.getText().toString();
                //创建Person对象
                Person person = new Person(name_str,paswd_str);
                //将Person对象保存为XML格式
                try{
                    //xml文件生成器
                    XmlSerializer serializer = Xml.newSerializer();
                    File file = new File(Environment.getExternalStorageDirectory(),"person_mes.xml");
                    FileOutputStream fi_out = new FileOutputStream(file);
                    serializer.setOutput(fi_out,"UTF-8");
                    serializer.startDocument("UTF-8",true);
                    serializer.startTag(null,"persons");
                    serializer.startTag(null,"person");
                    //写入Person对象的用户名属性
                    serializer.startTag(null,"name");
                    serializer.text(person.getUsername());
                    serializer.endTag(null,"name");
                    //写入Person对象的密码
                    serializer.startTag(null,"password");
                    serializer.text(person.getPaswd());
                    serializer.endTag(null,"password");
                    //结束标签
                    serializer.endTag(null,"person");
                    serializer.endTag(null,"persons");
                    serializer.endDocument();
                    serializer.flush();
                    fi_out.close();
Toast.makeText(this, "XML序列化成功!", Toast.LENGTH_SHORT).show();
                }catch (Exception e){
                    e.printStackTrace();
Toast.makeText(this, "XML序列化失败!", Toast.LENGTH_SHORT).show();
                }
                break;
            default:
                break;
        }
    }
}
```

### 4. 添加权限

由于上述代码实现的是将数据保存在 SD 卡中的 person_mes.xml 文件中,所以需要在 AndroidManifest.xml 文件中添加权限,具体代码如下所示:

```
< uses - permission android:name = "android.permission.WRITE_EXTERNAL_STORAGE" />
```

### 5. 运行程序

在完成了代码的编写和权限的添加以后,就可以运行程序。在用户界面输入用户的账号和密码,然后单击"XML 文件序列化"按钮,如果 Toast 提示"序列化失败",异常信息显示 Permission denied,则需要按照 6.2.1 节中图 6-1 的操作,为模拟器打开当前项目的存储权限,这样才可以使用。如果出现如图 6-10 所示的界面,则表明程序成功实现了用户注册信息的序列化。

### 6. 检查 person_mes.xml 文件是否生成

打开 Device File Explorer 视图,在 sdcard 下可以看到 person_mes.xml 文件,如图 6-11 所示。

### 7. 打开 person_mes.xml 文件

选中 person_mes.xml 文件,双击将其打开,查看内容是否与输入的一致,具体内容如下所示:

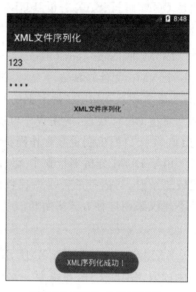

图 6-10　XML 文件序列化

图 6-11　person_mes.xml 文件

```
<?xml version = "1.0" encoding = "UTF-8" standalone = "true"?>
    < persons >
        < person >
            < name > 123 </name >
            < password > 1213 </password >
        </person >
    </persons >
```

出现上述内容就说明 XML 序列化成功。在使用 XML 序列化时,要注意节点的开始和结束、标签的开始和结束、字符的编码等。本节内容需要熟练掌握,因为使用的是类似于键值对的形式,所以在保存用户信息时经常使用。

### 6.3.3 XML 文件解析

在使用 XML 文件中的数据时,首先需要解析 XML 文件。通常解析 XML 文件有 3 种方式,分别是 DOM 解析、SAX 解析、PULL 解析。接下来将详细地介绍这 3 种方式。

**1. DOM 解析**

DOM 即 Document Object Model,文档对象模型。在应用程序中,基于 DOM 的 XML 分析器将一个 XML 文件转换成一个对象模型的集合(通常称 DOM 树),应用程序通过对这个对象模型的操作,来实现对 XML 文件数据的操作。

通过 DOM 接口,应用程序可以在任何时候访问 XML 文件中的任意一部分数据,因此,这种利用 DOM 接口的机制也被称作随机访问机制。

由于 DOM 分析器把整个 XML 文件转换成 DOM 树放在了内存中,因此,当文件比较大或结构比较复杂时,对内存的需求就比较高。所以较小的 XML 文件可以采用这种方式解析,但较大的文件不建议采用这种方式来解析。

**2. SAX 解析**

SAX(Simple API for XML,XML 简单应用程序接口)解析即事件驱动型的 XML 解析方式。顺序读取 XML 文件,不需要一次全部装载整个文件。当遇到像文件开头、文件结束或者标签开头与标签结束时,会触发一个事件,用户通过在其回调事件中写入处理代码来处理 XML 文件,适合对 XML 的顺序访问,且是只读的。移动设备的内存资源有限,所以 SAX 的顺序读取方式更适合移动开发。

**3. PULL 解析**

PULL 是 Android 内置的 XML 解析器。PULL 解析器的运行方式与 SAX 解析器相似,它提供了类似的事件,如:开始元素和结束元素事件,使用 parser.next()可以进入下一个元素并触发相应事件。事件将作为数值代码被发送,因此可以使用一个 switch 对感兴趣的事件进行处理。当元素开始解析时,调用 parser.nextText()方法可以获取下一个 Text 类型节点的值。

以上就是 XML 文件解析常用的 3 种方式,接下来将通过具体的案例讲解其中的一种——如何使用 PULL 解析 XML 文件。

### 6.3.4 XML 解析实例

视频讲解

在实际开发中,解析 XML 文件是常有的事,例如对保存的用户信息进行查看、天气预报信息展示等,所以掌握解析 XML 文件很重要。

使用 PULL 解析 XML 文件,首先要创建 XmlPullParser 解析器,通过该解析器提供的属性可以解析出 XML 文件中的各个节点的内容。常用属性如下所示:

XmlPullParser.START_DOCUMENT——XML 文件的开始,如<?xml version = "1.0" encoding="utf-8"?>。

XmlPullParser. END_DOCUMENT——XML 文件的结束。
XmlPullParser. START_TAG——XML 文件的开始节点。如<..>这种类型的。
XmlPullParser. END_TAG——XML 文件的结束节点。如<../>这种类型的。
PULL 解析器的具体使用步骤如下所示：
（1）调用 Xml. newPullParser()得到一个 XmlPullParser 对象。
（2）通过 parser. getEventType()获取到当前的事件类型。
（3）通过 while 循环判断当前的事件类型是否为文件结束。
（4）通过在 while 循环中使用的 switch 判断是否为开始标签，如果是，就获得标签的内容。

接下来将通过一个使用 PULL 解析用户信息的例子讲解 XML 文件的解析。PULL 解析的具体步骤如下。

### 1. 放入需要解析的 XML 文件

在 app 目录下创建一个 assets 文件夹，存放解析的文件，如图 6-12 所示。

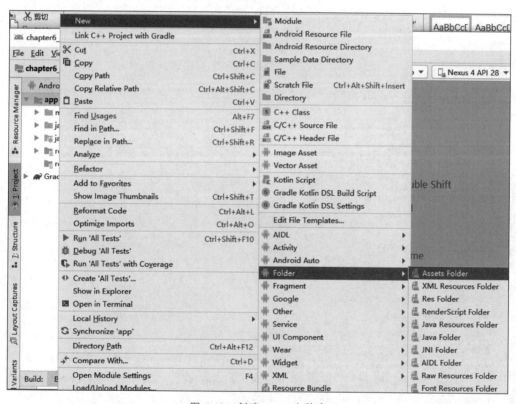

图 6-12　创建 assets 文件夹

把需要解析的 student. xml 文件放入 assets 文件夹下，student. xml 文件的具体代码如下所示：

```
<?xml version = "1.0" encoding = "utf-8"?>
<persons>
    <person id = "1">
        <name>liming</name>
```

```
            <age>23</age>
        </person>
        <person id="2">
            <name>zhangsan</name>
            <age>32</age>
        </person>
</persons>
```

## 2. 创建 Person 类

创建与上述 XML 文件中属性相对应的 Person 类，这个类封装了 ID、姓名、年龄。Person 类的代码如下所示：

```java
public class Person {
    private int id;
    private String name;
    private String age;
    public int getId() {
        return id;
    }
    public void setId(int id) {
        this.id = id;
    }
    public String getName() {
        return name;
    }
    public void setName(String name) {
        this.name = name;
    }
    public String getAge() {
        return age;
    }
    public void setAge(String age) {
        this.age = age;
    }
    @Override
    public String toString() {
        return "Person [id=" + id + ", name=" + name + ", age=" + age + "]";
    }
}
```

## 3. 创建 PersonService 类

为了更好地理解 PULL 解析 XML 文件的过程，特意采用将整个过程放在一个单独的 PersonService 类中，通过在 MainActivity 中调用类中定义的 getPersons()方法进行 XML 文件解析的方式。PersonService 类的具体代码如下所示：

```java
public class PersonService {
    public static List<Person> getPersons(InputStream xml) throws
IOException, XmlPullParserException {
        List<Person> persons = null;
        Person person = null;
        XmlPullParser pullParser = Xml.newPullParser();
        try {
            //为PULL解析器设置要解析的XML数据
            pullParser.setInput(xml, "UTF-8");
            int event = pullParser.getEventType();
            while(event!= XmlPullParser.END_DOCUMENT){
                switch (event) {
                    case XmlPullParser.START_DOCUMENT:
                        persons = new ArrayList<Person>();
                        break;
                    case XmlPullParser.START_TAG:
                        if("person".equals(pullParser.getName())){
                            int id = new Integer(pullParser.getAttributeValue(0));
                            person = new Person();
                            person.setId(id);
                        }
                        if("name".equals(pullParser.getName())){
                            String name = pullParser.nextText();
                            person.setName(name);
                        }
                        if("age".equals(pullParser.getName())){
                            String age = pullParser.nextText();
                            person.setAge(age);
                        }
                        break;
                    case XmlPullParser.END_TAG:
                        if("person".equals(pullParser.getName())){
                            persons.add(person);
                            person = null;
                        }
                        break;
                }
                **event = pullParser.next();**
            }
        } catch (XmlPullParserException e) {
            e.printStackTrace();
        }
        return persons;
    }
}
```

在上述代码中,需要注意的是 event=pullParser.next()这一句代码,因为在 while 循环中,当一个节点的信息解析完以后,会继续解析下一个节点,只有 type 的类型为 END_DOCUMENT 时才会结束循环,因此要把 pullParser.next()取到的值赋给 event。

### 4. 修改 activity_main.xml 布局代码

修改布局界面代码,布局中定义一个按钮"使用 PULL 解析 XML 文件"和一个文本区域,单击该按钮会在下方的文本区域显示解析到的内容。具体代码如下所示:

```xml
<?xml version = "1.0" encoding = "utf-8"?>
<LinearLayout xmlns:android = "http://schemas.android.com/apk/res/android"
    xmlns:app = "http://schemas.android.com/apk/res-auto"
    xmlns:tools = "http://schemas.android.com/tools"
    android:layout_width = "match_parent"
    android:layout_height = "match_parent"
    tools:context = "com.jxust.cn.chapter6_pullxml.MainActivity"
    android:orientation = "vertical">
    <Button
        android:id = "@+id/pulls"
        android:layout_width = "match_parent"
        android:layout_height = "wrap_content"
        android:text = "使用 PULL 解析 XML 文件" />
    <TextView
        android:id = "@+id/show_person"
        android:layout_marginLeft = "30dp"
        android:layout_marginTop = "50dp"
        android:layout_width = "wrap_content"
        android:layout_height = "wrap_content" />
</LinearLayout>
```

### 5. 修改 MainActivity 代码

因为在布局界面中定义了按钮和文本区域,所以需要在 Activity 中初始化按钮,然后为其添加监听。具体代码如下所示:

```java
public class MainActivity extends Activity {
    private Button button;
    private TextView show_person;
    @Override
    protected void onCreate(Bundle savedInstanceState) {
        super.onCreate(savedInstanceState);
        setContentView(R.layout.activity_main);
        show_person = (TextView)findViewById(R.id.show_person);
        button = (Button)findViewById(R.id.pulls);
        button.setOnClickListener(new View.OnClickListener() {
            @Override
            public void onClick(View view) {
                InputStream xml =
                    this.getClass().getClassLoader().getResourceAsStream("assets/person.xml");
                List<Person> persons = null;
```

```
                try {
                    persons = PersonService.getPersons(xml);
                } catch (IOException e) {
                    e.printStackTrace();
                } catch (XmlPullParserException e) {
                    e.printStackTrace();
                }
                for(Person person:persons){
                    show_person.setText(show_person.getText().toString() + "\n" + person.toString());
                }
            }
        });
    }
}
```

上述代码中,通过"InputStream xml = this.getClass().getClassLoader().getResourceAsStream("assets/person.xml");"这句代码取得文件,之后文件读取流对象,然后把获取到的信息以 TextView 文本的信息显示出来。

**6. 运行程序**

完成编码以后,运行程序,出现如图 6-13 所示的界面。

以上就是使用 PULL 解析 XML 文件的具体操作过程,由于 XML 文件经常的使用在服务器与客户端的数据交互中,所以需要对 XML 文件的生成以及解析熟练掌握与应用。

图 6-13 使用 PULL 解析 XML 文件

## 6.4 SharedPreferences 的使用

### 6.4.1 SharedPreferences 简介

SharedPreferences 类是一个轻量级的存储类,特别适合用于保存软件配置参数,例如用于登录时的用户名、密码、性别等参数。SharedPreferences 保存数据,其背后是用 XML 文件存放数据,文件存放在/data/data/<package name>/shared_prefs 目录下。

使用 SharedPreferences 需要先了解以下几个方法。

context.getSharedPreferences(name,mode):获取 SharedPreferences 的实例对象,方法的第一个参数用于指定该文件的名称,名称不用带后缀,后缀会由 Android 自动加上;方法的第二个参数指定文件的操作模式,共有 4 种操作模式。4 种操作模式分别为:

(1) MODE_APPEND——追加方式存储。

(2) MODE_PRIVATE——私有方式存储,其他应用无法访问。

(3) MODE_WORLD_READABLE——表示当前文件可以被其他应用读取。

(4) MODE_WORLD_WRITEABLE——表示当前文件可以被其他应用写入。

edit()方法：获取 Editor 对象，Editor 存储对象采用键值对进行存放。

commit()方法：提交数据。

上面讲解了使用 SharedPreferences 中需要用到的方法，接下来将分别通过一段示例代码讲解使用 SharedPreferences 存储数据、读取数据以及删除数据。

### 1. 使用 SharedPreferences 存储数据

使用 SharedPreferences 存储数据时，首先需要获取 SharedPreferences 对象，然后通过该对象获取到 Editor 对象，最后调用 Editor 对象的相关方法存储数据。具体代码如下所示：

```
SharedPreferences sharedPreferences = getSharedPreferences("test",
Context.MODE_PRIVATE);                          //私有数据
Editor editor = sharedPreferences.edit();        //获取编辑器
editor.putString("name", "江西");                //存入数据
editor.putString("history", "yejin");
editor.commit();                                 //提交修改
```

### 2. 使用 SharedPreferences 读取数据

使用 SharedPreferences 读取数据时，代码写起来相对较少，只需要创建 SharedPreferences 对象，然后使用该对象从对应的 key 取值即可。具体代码如下所示：

```
SharedPreferences sharedPreferences = context.getSharedPreferences();    //获取实例对象
String name = sharedPreferences.getString("name");                        //获取名字
String history = sharedPreferences.getString("history");                  //获取历史
```

### 3. 使用 SharedPreferences 删除数据

使用 SharedPreferences 删除数据时，首先需要获取到 Editor 对象，然后调用该对象的 remove()方法或者 clear()方法删除数据，最后提交。具体代码如下所示：

```
SharedPreferences sharedPreferences = context.getSharedPreferences();    //获取实例对象
Editor editor = sharedPreferences.edit();                                 //获取编辑器
editor.remove("name");                                                    //删除一条数据
editor.clear();                                                           //删除所有数据
editor.commit();                                                          //提交修改
```

以上就是使用 SharedPreferences 存储数据、删除数据以及读取数据的步骤。总结一下，就是在使用 SharedPreferences 存储数据和删除数据时需要使用 editor.commit()提交数据，取值时注意 key 的正确性即可。为了更好地学习 SharedPreferences 的使用，接下来将通过一个存储用户注册信息的例子来讲解。

视频讲解

## 6.4.2 使用 SharedPreferences 存储用户注册信息

6.4.1 节学习了 SharedPreferences 的简介、存储数据、读取数据以及删除数据的内容，下面将讲解如何使用 SharedPreferences 存储用户注册信息。

## 1. 创建用户注册信息的界面

首先需要设计用户注册的界面,修改 activity_main.xml 文件代码,实现如图 6-14 所示的用户注册界面。

图 6-14　用户注册界面

上述界面采用了线性布局的方式,先设置了一个 ImageView 组件显示头像,然后放了两个 EditText 和两个按钮,最后写了自定义样式的文件来设置 EditText 的样式。

activity_main.xml 文件的具体代码如下所示:

```xml
<?xml version = "1.0" encoding = "utf-8"?>
< LinearLayout xmlns:android = "http://schemas.android.com/apk/res/android"
    xmlns:app = "http://schemas.android.com/apk/res-auto"
    xmlns:tools = "http://schemas.android.com/tools"
    android:layout_width = "match_parent"
    android:layout_height = "match_parent"
    tools:context = "com.jxust.cn.chapter6_share.MainActivity"
    android:orientation = "vertical">
    < LinearLayout
        android:layout_width = "match_parent"
        android:layout_height = "wrap_content"
        android:layout_marginTop = "60dp"
        android:orientation = "vertical">
        < ImageView
            android:layout_gravity = "center_horizontal"
            android:layout_width = "70dp"
            android:layout_height = "70dp"
            android:background = "@drawable/icon"/>
    </LinearLayout >
    < LinearLayout
        android:layout_width = "match_parent"
        android:layout_height = "wrap_content"
        android:orientation = "vertical">
```

```xml
<EditText
    android:id="@+id/username"
    android:layout_marginTop="20dp"
    android:layout_width="match_parent"
    android:background="@drawable/edit_shape"
    android:hint="用户名"
    android:gravity="center"
    android:layout_height="40dp" />
<EditText
    android:id="@+id/paswd"
    android:layout_marginTop="20dp"
    android:background="@drawable/edit_shape"
    android:layout_width="match_parent"
    android:inputType="numberPassword"
    android:gravity="center"
    android:hint="密码"
    android:layout_height="40dp" />
<Button
    android:id="@+id/register"
    android:layout_width="match_parent"
    android:layout_height="wrap_content"
    android:textSize="18sp"
    android:text="注    册"
    android:layout_marginTop="20dp"
    android:background="#3F51B5"
    android:textColor="#FFFFFF"/>
<Button
    android:id="@+id/show"
    android:layout_width="match_parent"
    android:layout_height="wrap_content"
    android:textSize="18sp"
    android:text="显示保存的信息"
    android:layout_marginTop="20dp"
    android:background="#3F51B5"
    android:textColor="#FFFFFF"/>
    </LinearLayout>
</LinearLayout>
```

放在 drawable 目录下的自定义样式布局文件 edit_shape.xml 代码如下所示：

```xml
<?xml version="1.0" encoding="UTF-8"?>
<shape xmlns:android="http://schemas.android.com/apk/res/android">
    <solid android:color="#FFFFFF" />
    <corners android:radius="5dip"/>
    <stroke
        android:width="1dip"
        android:color="#728ea3" />
</shape>
```

## 2. 编写 MainActivity 代码

编写 MainActivity 中的代码，首先对两个 EditText 和两个按钮组件进行初始化，然后对按钮添加监听事件。第一个按钮的监听事件负责把用户输入的用户名和密码保存在 user_mes.xml 文件中；第二个按钮的监听事件负责把数据从文件中读取出来，然后显示给用户。具体代码如下所示：

```java
public class MainActivity extends Activity implements View.OnClickListener {
    private EditText username, paswd;
    private Button register_btn, show_btn;
    @Override
    protected void onCreate(Bundle savedInstanceState) {
        super.onCreate(savedInstanceState);
        setContentView(R.layout.activity_main);
        init();
    }
    //组件初始化方法
    private void init(){
        username = (EditText)findViewById(R.id.username);
        paswd = (EditText)findViewById(R.id.paswd);
        register_btn = (Button)findViewById(R.id.register);
        show_btn = (Button)findViewById(R.id.show);
        //对按钮添加监听
        register_btn.setOnClickListener(this);
        show_btn.setOnClickListener(this);
    }
    @Override
    public void onClick(View view) {
        switch (view.getId()){
            case R.id.register:
                //获取输入的用户名和密码
                String name_str = username.getText().toString();
                String paswd_str = paswd.getText().toString();
                boolean flag = save_userMes(MainActivity.this, name_str, paswd_str);
                if (flag){
                    Toast.makeText(MainActivity.this,"保存成功", Toast.LENGTH_SHORT).show();
                }else {
                    Toast.makeText(MainActivity.this,"保存失败", Toast.LENGTH_SHORT).show();
                }
                break;
            case R.id.show:
                Map<String, String> user = getuser_mes(MainActivity.this);
                if (user!= null){
                    String username = user.get("username");
                    String paswd = user.get("paswd");
                    Toast.makeText(MainActivity.this,"用户名为:" + username + "\n" + "密码为:" + paswd,
                            Toast.LENGTH_SHORT).show();
                }
                break;
            default:
                break;
```

```java
    }
}
//保存用户名和密码,并且生成 user_mes.xml 文件的方法
private boolean save_userMes(Context context, String username, String paswd) {
    SharedPreferences sharedPreferences = context.getSharedPreferences("user_mes", MODE_PRIVATE);
    SharedPreferences.Editor editor = sharedPreferences.edit();
    editor.putString("username", username);
    editor.putString("paswd", paswd);
    editor.commit();
    return true;
}
//从 user_mes.xml 文件中取出用户名和密码的方法
private Map<String, String> getuser_mes(Context context){
    SharedPreferences sharedPreferences =
            context.getSharedPreferences("user_mes", MODE_PRIVATE);
    String username = sharedPreferences.getString("username", null);
    String paswd = sharedPreferences.getString("paswd", null);
    Map<String,String> user = new HashMap<String, String>();
    user.put("username", username);
    user.put("paswd", paswd);
    return user;
}
```

上述代码包含了两个方法:一个是 save_userMes(),用来保存用户输入的信息;另外一个是 getuser_mes(),添加了验证,判断是否获取到用户信息,如果获取到则显示给用户。

### 3. 运行程序

完成以上编码以后,运行程序,首先输入用户名和密码,然后单击"注册"按钮,看是否提示用户,运行结果如图 6-15 所示;最后单击"显示保存的信息"按钮,看是否有信息显示,运行结果如图 6-16 所示。

图 6-15 "注册"结果图

图 6-16 "显示保存的信息"结果图

### 4. 查看是否生成了 user_mes.xml 文件

单击 Device File Explorer，找到 data/data/com.jxust.cn.chapter6_share/shared_prefs/目录下，查看的结果如图 6-17 所示。

图 6-17　user_mes.xml 文件生成图

### 5. 查看 user_mes.xml 文件内容

鼠标双击 user_mes.xml 文件，打开查看其中内容。具体内容如下所示：

```
<?xml version = '1.0' encoding = 'utf-8' standalone = 'yes' ?>
< map >
    < string name = "username"> 123 </string >
    < string name = "paswd"> 12312 </string >
</map >
```

以上就是使用 SharedPreferences 存储用户注册信息案例的操作过程，使用这种方式代码量相对较少，操作较简单，需要开发者熟练掌握。

## 6.5　SQLite 数据库

### 6.5.1　SQLite 数据库简介

SQLite 是一款轻型的数据库，是遵守 ACID 的关系型数据库管理系统，它包含在一个相对小的 C 库中。它是 D.RichardHipp 建立的公有领域项目，其设计目标是嵌入式的，而且目前已经在很多嵌入式产品中得到应用，它占用资源非常少，在嵌入式设备中，可能只需要几十万兆内存就够了。它能够支持 Windows/Linux/UNIX 等主流的操作系统，同时能够跟很多程序语言相结合，比如 C、C++、PHP、Perl、Java、C♯、Python、Ruby 等，还有 ODBC 接口，比起 MySQL、PostgreSQL 这两款世界著名的开源数据库管理系统，它的处理速度比它们都快。SQLite 第一个 Alpha 版本诞生于 2000 年 5 月，目前版本为 SQLite 3。

SQLite 数据库的工作原理如下。

不像常见的客户/服务器(C/S)范例，SQLite 引擎不是一个程序与之通信的独立进程，而是连接到程序中成为它的一个主要部分，所以主要的通信协议是在编程语言内直接 API 调用。这在消耗总量、延迟时间和整体简单性上有积极的作用。整个数据库(定义、表、索引和数据本身)都存放在宿主主机上，以一个单一文件的形式存储。

### 6.5.2　SQLite 数据库操作类以及接口

为了在 Android 中更方便地使用 SQLite 数据库，Android SDK 提供了许多对数据库操作的类和接口,接下来将针对常用的类及接口做详细的介绍。

### 1. SQLiteOpenHelper 类

SQLiteOpenHelper 类是 SQLiteDatabase 的一个帮助类,用来管理数据的创建和版本更新。一般的用法是定义一个类继承 SQLiteOpenHelper,并实现两个回调方法——OnCreate(SQLiteDatabase db)和 onUpgrade(SQLiteDatabse,int oldVersion,int newVersion)来创建和更新数据库。

onCreate()方法在初次生成数据库时才会被调用,在 onCreate()方法中可以生成数据库表结构及添加一些应用使用到的初始化数据。

onUpgrade()方法在数据库的版本发生变化时会被调用,一般在软件升级时才需改变版本号,而数据库的版本是由程序员控制的。

当程序调用 SQLiteOpenHelper 类的 getWritableDatabase()方法或者 getReadableDatabase()方法获取用于操作数据库的 SQLiteDatabase 实例的时候,如果数据库不存在,则 Android 系统会自动生成一个数据库,接着调用 onCreate()方法。

### 2. SQLiteDatabase 类

Android 提供了一个名为 SQLiteDatabase 的类(SQLiteOpenHelper 类中的 getWritableDatabase()和 getReadableDatabase()方法返回这个类的对象),使用该类可以完成对数据进行添加(Create)、查询(Retrieve)、更新(Update)和删除(Delete)操作(这些操作简称为 CRUD)。

SQLiteDatabase 类的常用方法如下所示。

- long insert(String table,String nullColumnHack,ContentValues values)——table:代表想插入数据的表名;nullColumnHack:代表强行插入 null 值的数据列的列名;values:代表一行记录的数据。insert()方法插入的一行记录使用 ContentValues 存放,ContentValues 类似于 Map,它提供了 put(String key,Xxx value)(其中 key 为数据列的列名)方法用于存入数据、getAsXxx(String key)方法用于取出数据。
- update(String table,ContentValues values,String whereClause,String[] whereArgs)——修改满足条件的数据。
- delete(String table,String whereClause,String[] whereArgs)——删除满足特定条件的数据。
- excelSQL(String sql,Object[] bindArgs)——执行一条占位符 SQL 语句。
- close()——关闭数据库。
- Cursor query(boolean distinct,String table,String[] columns,String selection,String[] selectionArgs,String groupBy,String having,String orderBy,String limit)——用于查询数据。

### 3. Cursor 接口

Cursor 是一个游标接口,在数据库中作为返回值,相当于结果集 ResultSet。Cursor 的一些常用方法如下所示:

- moveToFirst()——移动光标到第一行。
- moveToLast()——移动光标到最后一行。
- moveToNext()——移动光标到下一行。

- moveToPosition(int position)——移动光标到一个绝对的位置。
- moveToPrevious()——移动光标到上一行。
- getColumnCount()——返回所有列的总数。
- getColumnIndex(String columnName)——返回指定列的名称,如果不存在,则返回-1。
- getColumnName(int columnIndex)——从给定的索引返回列名。
- getColumnNames()——返回一个字符串数组的列名。
- getCount()——返回 Cursor 中的行数。

以上就是 SQLite 数据库常用的操作类以及接口,熟练掌握这些能够更好地进行程序中数据库的开发。

### 6.5.3 SQLite 数据库的操作

前面讲解了操作 SQLite 数据库常用的类及接口,接下来就通过这些类和接口来操作数据库,例如,数据库的创建、数据的添加、数据的删除、数据的更改等。下面将详细的讲解这几种操作的过程。

**1. SQLite 数据库的创建**

Android 系统在使用 SQLite 数据库时,一般使用 SQLiteOpenHelper 的子类创建 SQLite 数据库,因此需要创建一个类继承自 SQLiteOpenHelper,然后重写 onCreate()方法,在方法中执行创建数据库的语句。具体代码如下所示:

```java
public class User_Database extends SQLiteOpenHelper {
    public User_Database(Context context) {
        super(context, "user_db", null, 1);
    }
    //数据库第一次创建时调用该方法
    @Override
    public void onCreate(SQLiteDatabase sqLiteDatabase) {
        String sql = "create table user(id integer primary key autoincrement, username varchar(20),
       paswd varchar(20))";            //数据库执行语句
        sqLiteDatabase.execSQL(sql);
    }
    //数据库版本号更新时调用
    @Override
    public void onUpgrade(SQLiteDatabase sqLiteDatabase, int i, int i1) {
    }
}
```

上述代码创建了一个 user_db 数据库,然后通过 onCreate()方法创建了一个 user 表。里面有 id、username、paswd 属性。当数据库的版本更新时才会调用 onUpgrade()方法,如果版本没改变,则不调用该方法。MainActivity 中调用该方法在 data/data/com.jxust.cn.chapter6_sqlite/databases/目录下创建数据库的代码如下所示:

```
public class MainActivity extends AppCompatActivity {
    @Override
    protected void onCreate(Bundle savedInstanceState) {
        super.onCreate(savedInstanceState);
        setContentView(R.layout.activity_main);
        User_Database user = new User_Database(MainActivity.this);
        /* 只有调用 getReadableDatabase()或者 getWritableDatabase()函数后才能返回一个
           SQLiteDatabase 对象 */
        SQLiteDatabase sqLiteDatabase = user.getReadableDatabase();
    }
}
```

MainActivity 类调用执行完成以后会产生一个 user_db 数据库文件，如图 6-18 所示。

图 6-18  数据库文件

关于 SQLite 创建的数据库可以通过 SQLite Expert Professional 软件来查看。SQLite Expert Professional 是一款可视化 SQLite 数据库管理工具，SQLite Expert 允许用户在 SQLite 服务器上执行创建、编辑、复制、提取等操作。SQLite Expert Professional 支持所有的图形界面的 SQLite 特征。它包括一个可视化查询生成器、一个 SQL 编辑与语法突出和代码自动完成功能、强大的 table 和 view 设计与导入导出功能。开发者可以通过这款可视化操作软件查看和编辑创建的数据库，下载链接为 http://www.sqliteexpert.com/download.html。

### 2. 数据的添加

完成了数据库的创建，接下来就需要添加数据。添加数据的时候，首先需要获取一个 SQLiteDatabase 对象，在 User_Database 类中添加一个 adddata()方法，该方法用于添加数据。具体代码如下所示：

```
//添加数据
public void adddata(SQLiteDatabase sqLiteDatabase){
    ContentValues values = new ContentValues();
    values.put("username", "张三");
    values.put("paswd", "12222");
    sqLiteDatabase.insert("user", null, values);
    sqLiteDatabase.close();
}
```

MainActivity 调用该方法的代码如下所示：

```
SQLiteDatabase sql_date = user.getWritableDatabase();
user.adddata(sql_date);
```

最后把添加完数据以后的 user_db 文件保存到桌面,用 SQLite Expert Professional 软件打开,查看是否出现添加的数据,打开以后的内容如图 6-19 所示。

图 6-19　数据添加结果

以上就是使用 SQLite 数据库插入数据的操作。从图 6-19 可以看出,user 表中多了一条数据,与代码中插入语句的数据一致,说明数据插入成功。

### 3. 数据的删除

数据的删除首先也需要获取一个可写的 SQLiteDatabase 对象,在 User_Database 类中添加一个 delete()方法。具体代码如下所示:

```
//删除数据方法
public void delete(SQLiteDatabase sqLiteDatabase){
/*第一个参数:表名;第二个参数:需要删除的属性名,?代表占位符;第三个参数:属性名的属性值*/
sqLiteDatabase.delete("user", "username = ?", new String[]{"张三"});
sqLiteDatabase.close();}
```

以上就是数据的删除方法,在 MainActivity 中调用的方法和数据的添加是一样的。这里删除方法使用一个字符串和一个字符串数组来说明要删除的参数名和参数的值,删除后的结果如图 6-20 所示。

图 6-20　数据删除结果

### 4. 数据的更新

数据更新时使用 SQLiteDatabase 的 update()方法来修改表中的数据,也是首先需要先获取一个可写的 SQLiteDatabase 对象。具体代码如下所示:

```
//更新数据
public void update(SQLiteDatabase sqLiteDatabase){
    //创建一个 ContentValues 对象
    ContentValues values = new ContentValues();
    //以键值对的形式插入
    values.put("paswd","22233333");
    //执行修改的方法(修改 username = "张三"的密码)
```

```
        sqLiteDatabase.update("user",values,"username = ?",new String[]{"张三"});
        sqLiteDatabase.close();
}
```

上述就是把 user 表中 username＝"张三"的密码更新为 22233333 的具体操作,查看数据库文件的内容如图 6-21 所示。

图 6-21　数据的更新结果

以上就是数据库的更新操作的具体过程。在使用 SQLiteDatabase 对象后,要记得关闭数据库,否则会一直消耗内存,并且会报出数据库未关闭异常等。

### 6.5.4　使用 SQLite 数据库展示用户信息

视频讲解

6.5.3 节学习了数据库的增加、删除、查询、修改操作,接下来将通过一个具体的案例了解使用 SQLite 数据库实现用户信息的添加、删除、修改、查询等操作。用户的信息使用 ListView 来显示,接下来将详细讲解这些功能如何实现。

**1. 主界面布局设计(activity_main.xml)**

主界面布局中放置 3 个按钮,分别是"查询/删除用户信息""修改用户信息""添加用户信息",单击按钮后分别出现对应的界面。具体代码如下所示:

```
<?xml version = "1.0" encoding = "utf-8"?>
<LinearLayout xmlns:android = "http://schemas.android.com/apk/res/android"
    android:layout_width = "match_parent"
    android:layout_height = "match_parent"
    tools:context = "com.jxust.cn.chapter6_sqlite.MainActivity"
    android:orientation = "vertical">
    <Button
        android:id = "@ + id/search_delete"
        android:layout_width = "match_parent"
        android:layout_height = "wrap_content"
        android:layout_marginTop = "20dp"
        android:textSize = "17sp"
        android:textColor = "#FFFFFF"
        android:background = "#4169E1"
        android:text = "查询/删除用户信息" />
    <Button
        android:id = "@ + id/update"
        android:layout_width = "match_parent"
```

```
            android:layout_height = "wrap_content"
            android:layout_marginTop = "20dp"
            android:textSize = "17sp"
            android:textColor = "#FFFFFF"
            android:background = "#4169E1"
            android:text = "修改用户信息" />
    <Button
            android:id = "@+id/add"
            android:layout_width = "match_parent"
            android:layout_height = "wrap_content"
            android:layout_marginTop = "20dp"
            android:textSize = "17sp"
            android:textColor = "#FFFFFF"
            android:background = "#4169E1"
            android:text = "添加用户信息" />
</LinearLayout>
```

上述代码中使采用线性垂直的布局方式定义了 3 个按钮,实现的界面如图 6-22 所示。

图 6-22　主界面

### 2. 数据库创建以及操作方法类

数据库的创建类 User_Database 包含了数据库创建的 onCreate()、版本更新的 onUpgrade()、数据添加的 adddata()、数据删除的 delete()以及数据更新的 update()方法,具体代码如下所示:

```
public class User_Database extends SQLiteOpenHelper {
    public User_Database(Context context) {
        super(context, "user_db",null,1);
    }
    //数据库第一次创建时调用该方法
    @Override
    public void onCreate(SQLiteDatabase sqLiteDatabase) {
        //数据库执行语句
        String sql = "create table user(id integer primary key autoincrement,
username varchar(20),paswd varchar(20),sex varchar(20),age integer)";
```

```java
        sqLiteDatabase.execSQL(sql);
}
//数据库版本号更新时调用
@Override
public void onUpgrade(SQLiteDatabase sqLiteDatabase, int i, int i1) {
}
//添加数据
public void adddata(SQLiteDatabase sqLiteDatabase, String username, String paswd, String sex, int age){
    ContentValues values = new ContentValues();
    values.put("username", username);
    values.put("paswd", paswd);
    values.put("sex", sex);
    values.put("age", age);
    sqLiteDatabase.insert("user", null, values);
    sqLiteDatabase.close();
}
//删除数据
public void delete(SQLiteDatabase sqLiteDatabase, int id){
    /* 第一个参数:表名;第二个参数:需要删除的属性名,?代表占位符;第三个参数:属性名的属
       性值 */
    sqLiteDatabase.delete("user", "id = ?", new String[]{id + ""});
    sqLiteDatabase.close();
}
//更新数据
public void update(SQLiteDatabase sqLiteDatabase, int id, String username, String paswd, String
                sex, int age){
    //创建一个ContentValues对象
    ContentValues values = new ContentValues();
    //以键值对的形式插入
    values.put("username", username);
    values.put("paswd", paswd);
    values.put("sex", sex);
    values.put("age", age);
    //执行修改的方法
    sqLiteDatabase.update("user", values, "id = ?", new String[]{id + ""});
    sqLiteDatabase.close();
}
//查询数据
public List<UserInfo> querydata(SQLiteDatabase sqLiteDatabase){
    Cursor cursor = sqLiteDatabase.query("user", null, null, null, null, null, "id ASC");
    List<UserInfo> list = new ArrayList<UserInfo>();
    while(cursor.moveToNext()){
        int id = cursor.getInt(cursor.getColumnIndex("id"));
        String username = cursor.getString(1);
        String paswd = cursor.getString(2);
        String sex = cursor.getString(3);
        int age = cursor.getInt(cursor.getColumnIndex("age"));
        list.add(new userInfo(id, username, paswd, sex, age));
```

```
        }
        cursor.close();
        sqLiteDatabase.close();
        return list;
    }
}
```

### 3. 创建一个 UserInfo 的 JavaBean 类

在操作数据库时,把数据存放在一个 JavaBean 对象中操作起来会相对简单一点,因此创建一个 UserInfo 类。具体代码如下所示:

```
public class UserInfo {
    private int id;
    private String username;
    private String paswd;
    private String sex;
    private int age;
    //构造方法
    public UserInfo(int id, String username, String paswd, String sex, int age) {
        this.id = id;
        this.username = username;
        this.paswd = paswd;
        this.sex = sex;
        this.age = age;
    }
    public void setId(int id) {
        this.id = id;
    }
    public void setUsername(String username) {
        this.username = username;
    }
    public void setPaswd(String paswd) {
        this.paswd = paswd;
    }
    public void setSex(String sex) {
        this.sex = sex;
    }
    public void setAge(int age) {
        this.age = age;
    }
    public int getId() {
        return id;
    }
    public String getUsername() {
        return username;
    }
    public String getPaswd() {
```

```
            return paswd;
    }
    public String getSex() {
        return sex;
    }
    public int getAge() {
        return age;
    }
    @Override
    public String toString() {
        return "userInfo{" + "id = " + id + ", username = '" + username + '\'' +", paswd = '"
                    + paswd + '\'' + ", sex = '" + sex + '\'' +", age = " + age + '}';
    }
}
```

### 4. 主界面功能实现类（MainActivity）

MainActivity 负责数据库的创建以及对布局中的 3 个按钮添加监听,然后跳转到不同的功能界面中去。具体代码如下所示：

```
public class MainActivity extends Activity implements View.OnClickListener {
    public User_Database user;
    public SQLiteDatabase sqL_read;
    private Button search_del_btn, insert_btn, update_btn;
    @Override
    protected void onCreate(Bundle savedInstanceState) {
        super.onCreate(savedInstanceState);
        setContentView(R.layout.activity_main);
        User_Database user = new User_Database(MainActivity.this);
        //获取一个可读的数据库
        sqL_read = user.getReadableDatabase();
        init();
    }
    //组件初始化方法
    public void init(){
        search_del_btn = (Button)findViewById(R.id.search_delete);
        insert_btn = (Button)findViewById(R.id.add);
        update_btn = (Button)findViewById(R.id.update);
        //添加监听
        search_del_btn.setOnClickListener(this);
        insert_btn.setOnClickListener(this);
        update_btn.setOnClickListener(this);
    }
    @Override
    public void onClick(View view) {
        switch (view.getId()){
            case R.id.search_delete:
                Intent intent1 = new Intent(MainActivity.this, Sea_deluser_Activity.class);
```

```
                startActivity(intent1);
                break;
            case R.id.add:
                Intent intent2 = new Intent(MainActivity.this, Insertuser_Activity.class);
                startActivity(intent2);
                break;
            case R.id.update:
                Intent intent3 = new Intent(MainActivity.this, Updareuser_Activity.class);
                startActivity(intent3);
                break;
            default:
                break;
        }
    }
}
```

上述程序运行以后,可以看到在项目的目录下产生了一个名为 user_db 的文件,为了下面查询与删除操作功能的实现,先手动在数据库中添加几条数据,具体如图 6-23 所示。

图 6-23 手动添加数据

### 5．用户信息查询与删除界面设计(user_search_delete.xml 与 detail.xml)

考虑到删除用户信息,需要先查询到用户信息,然后再根据 ID 删除数据,所以把用户的信息查询与删除放在一个界面里实现。在显示用户的信息时,单击用户名会弹出一个对话框,提示用户是否删除。其中 user_search_delete.xml 的具体代码如下所示:

```xml
<?xml version = "1.0" encoding = "utf-8"?>
<LinearLayout xmlns:android = "http://schemas.android.com/apk/res/android"
    android:layout_width = "match_parent"
    android:layout_height = "match_parent"
    android:orientation = "vertical">
    <LinearLayout
        android:layout_width = "match_parent"
        android:layout_height = "50dp"
        android:background = "#3F51B5"
        android:orientation = "horizontal">
        <ImageView
            android:layout_width = "50dp"
```

```xml
                android:layout_height = "50dp"
                android:layout_marginLeft = "10dp"
                android:src = "@mipmap/ic_launcher"/>
        <TextView
                android:layout_width = "wrap_content"
                android:layout_height = "match_parent"
                android:text = "查询用户"
                android:textSize = "18sp"
                android:gravity = "center_vertical"
                android:textColor = "#FFFFFF"/>
    </LinearLayout>
    <ListView
        android:id = "@+id/mes"
        android:layout_width = "match_parent"
        android:dividerHeight = "2dp"
        android:layout_height = "wrap_content">
    </ListView>
</LinearLayout>
```

### 6. 用户信息查询 & 删除功能实现类（Sea_deluser_Activity）

具体代码如下所示：

```java
public class Sea_deluser_Activity extends Activity {
    public ListView user_list;
    private List<UserInfo> list;
    private SQLiteDatabase sqLiteDatabase;
    //假设数据库用户不超过10个
    private String[] user_mes;
    @Override
    protected void onCreate(Bundle savedInstanceState) {
        super.onCreate(savedInstanceState);
        requestWindowFeature(Window.FEATURE_NO_TITLE);
        setContentView(R.layout.user_search_delete);
        user_list = findViewById(R.id.mes);
        User_Database users = new User_Database(Sea_deluser_Activity.this);
        sqLiteDatabase = users.getReadableDatabase();
        //获取从数据库查询到的数据
        list = users.querydata(sqLiteDatabase);
        //把获取到的信息添加到用户名数组中
        user_mes = new String[list.size()];
        for(int i = 0;i<list.size();i++){
            user_mes[i] = list.get(i).getUsername() + "    " +
                    list.get(i).getPaswd() + "    " + list.get(i).getAge() + "    " +
                    list.get(i).getSex();
        }
        //把用户名显示在ListView上
        final ArrayAdapter<String> adapter = new ArrayAdapter<String>
```

```java
            (Sea_deluser_Activity.this, android.R.layout.simple_list_item_1, user_mes);
        user_list.setAdapter(adapter);
        //为ListView每个元素添加单击事件
        user_list.setOnItemClickListener(new AdapterView.OnItemClickListener() {
            @Override
            public void onItemClick(AdapterView<?> adapterView, View view, int I, long l)
            {
                final int id = list.get(i).getId();
                //弹出一个对话框
                new AlertDialog.Builder(Sea_deluser_Activity.this).setTitle("系统提示")
                        //设置显示的内容
                        .setMessage("确定删除该条数据吗?")
                        //添加"确定"按钮
                        .setPositiveButton("确定",new DialogInterface.OnClickListener() {
                            public void onClick(DialogInterface dialog, int which) {
                                //删除数据操作,首先获取到ID
                                User_Database user_database =
                                        new User_Database(Sea_deluser_Activity.this);
                                SQLiteDatabase sqLiteDatabase =
                                        user_database.getWritableDatabase();
                                user_database.delete(sqLiteDatabase, id);
                                refresh();
Toast.makeText(Sea_deluser_Activity.this, "删除成功", Toast.LENGTH_SHORT).show();
                        }).setNegativeButton("取消", new DialogInterface.OnClickListener() {
                            //添加"取消"按钮
                            @Override
                            public void onClick(DialogInterface dialog, int which) {
                            }
                        }).show(); //在按键响应事件中显示此对话框
            }
        });
    }
    //刷新页面方法
    private void refresh() {
        finish();
        Intent intent = new Intent(Sea_deluser_Activity.this, Sea_deluser_Activity.class);
        startActivity(intent);
    }
}
```

以上就是用户信息查询和删除的代码,需要注意的是,因为把数据库操作的方法放在一个单独的类中,每次SQLDatabase的对象都会把数据库关闭,所以需要重新创建一个SQLDatabase对象。以上代码的实现结果如图6-24所示。

### 7. 用户信息添加界面设计(user_insert.xml)

用户信息添加界面主要包含3个EditText组件,分别对应数据库中的username、paswd、age。另外还包括一个Spinner组件和一个按钮,Spinner组件负责让用户选择性别;按钮负责把数据添

图 6-24　查询、删除对话框、删除结果图

加到数据库中。user_insert.xml 文件的代码具体如下所示：

```xml
<?xml version = "1.0" encoding = "utf-8"?>
<LinearLayout xmlns:android = "http://schemas.android.com/apk/res/android"
    android:layout_width = "match_parent"
    android:layout_height = "match_parent"
    android:orientation = "vertical">
    <EditText
        android:id = "@ + id/insert_name"
        android:layout_width = "match_parent"
        android:layout_height = "40dp"
        android:gravity = "center"
        android:background = "@drawable/edit_shape"
        android:layout_marginTop = "10dp"
        android:hint = "请输入用户名"/>
    <EditText
        android:id = "@ + id/insert_paswd"
        android:layout_width = "match_parent"
        android:layout_height = "40dp"
        android:inputType = "numberPassword"
        android:gravity = "center"
        android:background = "@drawable/edit_shape"
        android:layout_marginTop = "10dp"
        android:hint = "请输入密码"/>
    <Spinner
        android:id = "@ + id/insert_sex"
        android:layout_width = "match_parent"
        android:layout_height = "wrap_content"
        android:gravity = "center"
        android:layout_marginTop = "10dp"
        android:entries = "@array/sex" />
    <EditText
        android:id = "@ + id/insert_age"
        android:layout_width = "match_parent"
        android:layout_height = "40dp"
```

```
            android:gravity = "center"
            android:background = "@drawable/edit_shape"
            android:hint = "请输入年龄"/>
    <Button
    android:id = "@ + id/save_usermes"
            android:layout_marginTop = "20dp"
            android:layout_width = "match_parent"
            android:textSize = "17sp"
            android:textColor = "#FFFFFF"
            android:background = "#4169E1"
            android:text = "添加该用户信息"
            android:layout_height = "50dp" />
</LinearLayout>
```

上述代码是插入用户信息的布局代码,实现的界面如图 6-25 所示。

图 6-25　添加用户信息

### 8. 用户信息添加功能实现类(Insertuser_Activity)

Insertuser_Activity 首先需要获取各个组件,然后为按钮添加监听事件,为 Spinner 组件添加选择事件。按钮的监听事件负责把 EdiitText 中输入的信息和 Spinner 中选择的性别添加到数据库中,然后跳转到用户数据查询页面,查看插入的信息,添加的事件通过调用数据库创建类的数据插入方法实现。具体的实现代码如下:

```
public class Insertuser_Activity extends Activity {
    private EditText name_edit, paswd_edit, age_edit;
    private Spinner spinner;
    private Button save_btn;
    private String select_sex = "男";
    @Override
    protected void onCreate(Bundle savedInstanceState) {
        super.onCreate(savedInstanceState);
        requestWindowFeature(Window.FEATURE_NO_TITLE);
        setContentView(R.layout.user_insert);
        init();
    }
    public void init(){
```

```java
        name_edit = (EditText)findViewById(R.id.insert_name);
        paswd_edit = (EditText)findViewById(R.id.insert_paswd);
        spinner = (Spinner) findViewById(R.id.insert_sex);
        //为选择性别下拉列表添加选择事件
        spinner.setOnItemSelectedListener(new AdapterView.OnItemSelectedListener() {
            @Override
        public void onItemSelected(AdapterView<?> adapterView, View view, int i, long l){
        //获取选择的值
        select_sex = Insertuser_Activity.this.getResources().getStringArray(R.array.sex)[i];
    }
            @Override
            public void onNothingSelected(AdapterView<?> adapterView) {
            }
        });
        age_edit = (EditText)findViewById(R.id.insert_age);
        save_btn = (Button)findViewById(R.id.save_usermes);
        save_btn.setOnClickListener(new View.OnClickListener() {
            @Override
            public void onClick(View view) {
                //获取用户输入的用户名、密码、年龄
                String name_str = name_edit.getText().toString();
                String paswd_str = paswd_edit.getText().toString();
                int age = Integer.parseInt(age_edit.getText().toString());
                //调用数据库操作类的插入方法
                User_Database us_db = new User_Database(Insertuser_Activity.this);
                SQLiteDatabase sqLiteDatabase = us_db.getWritableDatabase();
                us_db.adddata(sqLiteDatabase, name_str, paswd_str, select_sex, age);
                Intent intent = new Intent(Insertuser_Activity.this, Sea_deluser_Activity.class);
                startActivity(intent);
            }
        });
    }
}
```

以上就是添加数据的处理代码,因为性别的下拉列表里边的值是存放在 values/arrays 下的,所以通过 Insertuser_Activity.this.getResources().getStringArray(R.array.sex)[i]方法取得选择的值。在取得用户输入的用户名、密码、年龄以后,创建 User_Database 类实例,调用该实例的添加数据方法。添加数据操作以及添加完重新查询更新后的数据结果如图 6-26 所示。

以上就是添加数据的操作过程,从图 6-26 中可以看到,在左图中填写数据以后,单击"添加该用户信息"按钮,会跳转到"查询用户"界面,此时数据是从数据库重新获取到的,实现了数据的实时刷新。为了更加清楚地看到数据的变化,可以把数据库文件 user_db 导出到桌面上。user_db 打开后的具体内容如图 6-27 所示。

本节通过使用 SQLite 数据库管理用户信息的案例,讲解了如何创建数据库以及数据的添加、数据的删除、数据的查询、数据的更新操作。在这里没有写出用户信息更新的功能,可以按照用户删除的操作,使用对话框的形式来实现。

图 6-26　添加数据操作以及更新后的结果

图 6-27　数据库变化

## 6.6　JSON 学习

### 6.6.1　JSON 简介

JSON 的全称是 JavaScript Object Notation，它是一种轻量级的数据交换语言，以文字为基础，且易于让人阅读，同时也方便了机器进行解析和生成。

简单地说，JSON 就是 JavaScript 中的对象和数组，这两种结构就是对象和数组两种结构，通过这两种结构可以表示各种复杂的结构，其可以将 JavaScript 对象中表示的一组数据转换为字符串，然后可以在函数之间轻松地传递这个字符串，或者在异步应用程序中将字符串从 Web 客户机传递给服务器端程序。

JSON 采用完全独立于程序语言的文本格式，但是也使用了类 C 语言的习惯（包括 C、C++、C♯、Java、JavaScript、Perl、Python 等）。这些特性使 JSON 成为理想的数据交换语言。

**1. JSON 的基础结构**

（1）对象。对象在 JavaScript 中表示为"{ }"括起来的内容，数据结构为{key：value，key：value，…}的键值对形式，在面向对象的语言中，key 为对象的属性，value 为对应的属性值，所以很容易理解，取值方法为对象.key 获取属性值，这个属性值的类型可以是数字、字符串、数组、对象。

（2）数组。数组在 JavaScript 中是中括号"[ ]"括起来的内容，数据结构为 [ "Java"，"JavaScript"，"VB"，…]，取值方式和所有语言一样，使用索引获取，字段值的类型可以是数字、字符串、数组、对象。

## 2. 使用JSON语法创建对象

使用JSON语法创建对象是一种更简单的方式,它直接获取各JavaScript对象。对于以前旧版本的JavaScript来说,JSON使用它创建对象的方式如下所示:

```javascript
//定义一个函数
function user(name, age){
    this.name = name;
    this.age = age;
}
//创建一个user对象
var ur = new user("小明","12");
//输出user实例的年龄
alert(ur.name);
```

从JavaScript 1.2版本以后,创建对象有了一种更快捷的语法,代码如下所示:

```javascript
var ur = {"name":'小明',"age":'12'};
alert(ur);
```

上述语法就是一种JSON语法,显然使用JSON语法创建对象会更加简便。这种语法创建对象时,总以"{"开始,以"}"结束,对象的每个属性名和属性值之间用英文冒号(:)隔开,多个属性值之间用英文逗号(,)隔开。需要注意的是,只有当前属性值后面还有定义的属性时,才用逗号隔开。

## 3. 使用JSON语法创建数组

使用JSON语法创建数组也是非常常见的,在早期的JavaScript语法中,开发者通过如下所示的方式创建数组:

```javascript
//首先创建数组对象
var arr = new Array();
//为数组元素赋值
arr[0] = '1';
arr[1] = '2';
```

如果使用JSON语法,则可以通过如下方式创建数组:

```javascript
//使用JSON语法创建数组
var arr = ['1','2'];
```

从上述代码中可以看到,JSON创建数组总以"["开始,然后依次放入数组元素,元素与元素之间以英文逗号隔开,最后一个数组元素不需要逗号,最后以"]"结束。

## 4. Java对JSON的支持

当服务器返回一个满足JSON格式的字符串后,开发者可以利用项目提供的工具类将该字符串转换为JSON对象或JSON数组。

在Android系统中,内置了对JSON的支持,在Android SDK的orj.json包下提供了

JSONArray、JSONObject、JSONStringer 等类，通过这些类即可完成 JSON 字符串与 JSONArray、JSONObject 之间的转换。

Java 的 JSON 支持主要依赖于 JSONArray、JSONObject 两个类。其中 JSONArray 代表一个数组，它可完成 Java 集合（集合元素可以是对象）与 JSON 字符串之间的相互转换；JSONObject 代表一个 JSON 对象，它可完成 Java 对象与 JSON 字符串之间的相互转换。

## 6.6.2 JSON 解析案例

视频讲解

在了解了 JSON 创建对象、数组以及 Java 对 JSON 的支持以后，接下来将通过一个解析 JSON 数据格式文件的案例来讲解 JSON 解析的使用，分别采用 Android 和 Java 实现。

**1. 利用 Android 实现**

test.json 的解析方法具体步骤如下。

1）定义 test.json 文件

```
{
    "cat":"it",
    "languages":[
        {"id":1,"ide":"Eclipse","name":"Java"},
        {"id":2,"ide":"Xcode","name":"Swift"},
        {"id":3,"ide":"Visual Studio","name":"C#"}
    ],
    "pop":true
}
```

2）创建项目 Chapter6_ReadJson

与 6.3.4 节解析 XML 文件相同，先在 App 目录下创建一个 assets 文件夹，将 test.json 文件放到 assets 文件夹下，如图 6-28 所示。

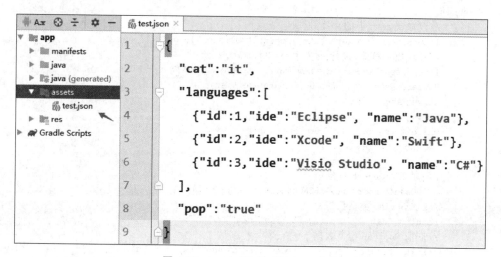

图 6-28　assets 下的 test.json 文件

3）修改 activity_main.xml 布局代码

修改布局界面代码，布局中定义一个文本区域居中显示，运行程序直接在文本区域显示解析到的内容。具体代码如下所示：

```xml
<LinearLayout xmlns:android="http://schemas.android.com/apk/res/android"
    xmlns:app="http://schemas.android.com/apk/res-auto"
    xmlns:tools="http://schemas.android.com/tools"
    android:layout_width="match_parent"
    android:layout_height="match_parent"
    tools:context=".MainActivity">
    <TextView
        android:layout_width="match_parent"
        android:layout_height="wrap_content"
        android:layout_gravity="center"
        android:gravity="center"
        android:text="Hello World!"
        android:id="@+id/tvShow"/>
</LinearLayout>
```

4）修改 MainActivity 代码

因为在布局界面中定义了文本区域，所以需要在 Activity 中初始化 TextView。具体代码如下所示：

```java
public class MainActivity extends AppCompatActivity {
    TextView tvShow;
    String result = "";
    @Override
    protected void onCreate(Bundle savedInstanceState) {
        super.onCreate(savedInstanceState);
        setContentView(R.layout.activity_main);
        tvShow = findViewById(R.id.tvShow);
        try {
            //读取 assets 下的文件，返回字节流
            InputStream is = getAssets().open("test.json");
            //转换成字符流，按行读取，存入 StringBuilder
            InputStreamReader isr = new InputStreamReader(is);
            BufferedReader br = new BufferedReader(isr);
            String str = "";
            StringBuilder sb = new StringBuilder();
            while((str = br.readLine())!= null){
                sb.append(str);
            }
            //创建 JSONObject 对象
            JSONObject root = new JSONObject(sb.toString());
            //获取 String 类型数据
            result = "cat = " + root.getString("cat") + "\n";
            //获取 Boolean 类型数据
            result = result + "pop = " + root.getBoolean("pop") + "\n";
            //获取 JSON 数组
```

```
            JSONArray array = root.getJSONArray("languages");
            for(int i = 0;i < array.length();i++) {
                result = result + " ---------------------- " + "\n";
                JSONObject lan = array.getJSONObject(i);
                String tmp = "id = " + lan.get("id") + "\n" + "ide = " + lan.get("ide") + "\n" +
                             "name = " + lan.get("name") + "\n";
                result = result + tmp;
            }
            tvShow.setText(result);
        } catch (IOException e) {
            e.printStackTrace();
        } catch (JSONException e) {
            e.printStackTrace();
        }
    }
}
```

5）运行程序

完成编码以后,运行程序,出现如图 6-29 所示的界面。

从图 6-29 可以看到,文件的内容解析正确,每一组值都以键值对的形式打印出来。Android 中集成了 org.json 包,解析 JSON 文件时引入 org.json.JSONObject 和 org.json.JSONArray,可分别创建 JsonObject 对象、得到 JSON 数组,然后根据键来取值。

**2. 利用 Java 实现 test.json 文件的解析方法**

（1）在 Eclipse for Java 中创建 test 项目,将 test.json 文件放在同级目录下,同时创建一个 Java 文件。下载解析需要使用到的 JSON 包,下载地址为 https://repo1.maven.org/maven2/com/google/code/gson/gson/2.8.1/,然后将下载的包导入到项目当中。项目的具体结构如图 6-30 所示。

图 6-29 解析 JSON 文件

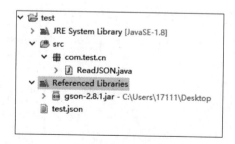

图 6-30 项目结构

（2）编写负责解析的 Java 类 ReadJSON.java。具体代码如下所示：

```
import java.io.FileNotFoundException;
import java.io.FileReader;
```

```java
import com.google.gson.JsonArray;
import com.google.gson.JsonIOException;
import com.google.gson.JsonObject;
import com.google.gson.JsonParser;
import com.google.gson.JsonSyntaxException;
public class ReadJSON {
    public static void main(String args[]){
        try {
            JsonParser parser = new JsonParser();          //创建 JSON 解析器
            //创建 JsonObject 对象
            JsonObject object = (JsonObject) parser.parse(new FileReader("test.json"));
            //将 JSON 数据转换为 String 类型的数据
            System.out.println("cat = " + object.get("cat").getAsString());
            //将 JSON 数据转换为 Boolean 类型的数据
            System.out.println("pop = " + object.get("pop").getAsBoolean());
            //得到 JSON 的数组
            JsonArray array = object.get("languages").getAsJsonArray();
            for(int i = 0;i < array.size();i++){
                System.out.println("----------------");
                JsonObject subObject = array.get(i).getAsJsonObject();
                System.out.println("id = " + subObject.get("id").getAsInt());
                System.out.println("name = " + subObject.get("name").getAsString());
                System.out.println("ide = " + subObject.get("ide").getAsString());
            }
        } catch (JsonIOException e) {
            e.printStackTrace();
        } catch (JsonSyntaxException e) {
            e.printStackTrace();
        } catch (FileNotFoundException e) {
            e.printStackTrace();
        }
    }
}
```

(3) 运行 ReadJSON.java 程序,在控制台会打印出与图 6-29 相同的结果,如图 6-31 所示。

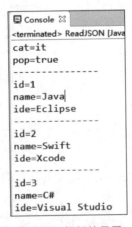

图 6-31　解析结果图

以上就是使用 Android 和 Java 解析 JSON 文件的具体操作步骤,开发者可以学习使用 Java 来生成 JSON 文件或者在后续的章节中通过网络请求与服务器进行 JSON 格式的数据交换。

## 6.7 【项目案例 5】 "橙一色"App 数据存储及主页展示

视频讲解

本章介绍了 Android 数据存储的几种方式。接下来"橙一色"App 使用 SQLite 数据库来实现用户注册信息保存、登录时从数据库查询验证的功能。定义 SQLite 数据库操作类,用来操作数据库的创建以及信息的查询等功能。即定义一个类 OrangeDatabase,继承自 SQLiteOpenHelper,然后重写其方法。创建一个名为 orange 的数据库,并在其下创建 orange_user 表。具体代码如下:

```java
public class OrangeDatabase extends SQLiteOpenHelper {
    public OrangeDatabase(@Nullable Context context) {
        super(context, "orange", null, 1);
    }
    @Override
    public void onCreate(SQLiteDatabase db) {
        //创建用户表
        String sql = "create table orange_user(id integer primary key autoincrement,
          username varchar(50), password varchar(50), sex varchar(10), city carchar(50))";
        db.execSQL(sql);
    }
    @Override
    public void onUpgrade(SQLiteDatabase db, int oldVersion, int newVersion) {
    }
    //插入数据
    public void insertUser(SQLiteDatabase sqLiteDatabase, String username, String password, String sex, String city) {
        ContentValues contentValues = new ContentValues();
        contentValues.put("username", username);
        contentValues.put("password", password);
        contentValues.put("sex", sex);
        contentValues.put("city", city);
        sqLiteDatabase.insert("orange_user", null, contentValues);
        sqLiteDatabase.close();
    }
    //查询数据
    public Bundle queryUserInfo(SQLiteDatabase sqLiteDatabase, Bundle bundle) {
        String username = bundle.getString("username");
        Cursor cursor = sqLiteDatabase.rawQuery("select * from orange_user where username = ?", new
                         String[]{username});
        if (cursor != null) {
            while (cursor.moveToNext()) {
                bundle.putString("sex", cursor.getString(3));
                bundle.putString("city", cursor.getString(4));
            }
        }
```

```
            cursor.close();
            sqLiteDatabase.close();
            return bundle;
    }
}
```

然后在 MainActivity 和 RegisterActivity 中进行调用。MainActivity 主要是在登录操作时进行验证,将输入的用户名与密码和数据库中存储的数据进行匹配,看是否有该用户,进而判断是否能够登录。具体代码如下:

```
public class MainActivity extends AppCompatActivity implements OnClickListener {
    private Button registerButton, loginButton;
    private EditText usernameText, paswdEdit;
    @Override
    protected void onCreate(Bundle savedInstanceState) {
        super.onCreate(savedInstanceState);
        setContentView(R.layout.user_login);
        init();
    }
    @Override
    public void onClick(View v) {
        switch (v.getId()) {
            case R.id.register:
                Intent intent = new Intent(MainActivity.this, RegisterActivity.class);
                startActivity(intent);
                break;
            case R.id.login:
                //注册时,引入了数据库,登录可以通过数据库进行验证,验证通过跳转到首页,不通过
                //进行提示
                if (validateLogin()) {
                    Intent intent1 = newIntent(MainActivity.this, IndexActivity.class);
                    Bundle bundle = new Bundle();
        OrangeDatabase orangeDatabase = new OrangeDatabase(MainActivity.this);
                    bundle.putString("username",usernameText.getText().toString());
                    bundle = orangeDatabase.queryUserInfo(
                        orangeDatabase.getReadableDatabase(), bundle);
                    intent1.putExtras(bundle);
                    startActivity(intent1);
                } else {
    Toast.makeText(MainActivity.this, "账号或者密码错误", Toast.LENGTH_SHORT).show();
                }
                break;
        }
    }
    //界面组件初始化
    private void init() {
        usernameText = findViewById(R.id.username);
```

```java
        paswdEdit = findViewById(R.id.password);
        loginButton = findViewById(R.id.login);
        loginButton.setOnClickListener(this);
        registerButton = findViewById(R.id.register);
        registerButton.setOnClickListener(this);
    }
    //登录验证
    private boolean validateLogin() {
        String username = usernameText.getText().toString();
        String password = paswdEdit.getText().toString();
        OrangeDatabase orangeDatabase = new OrangeDatabase(MainActivity.this);
        SQLiteDatabase sqLiteDatabase = orangeDatabase.getReadableDatabase();
        Cursor cursor = sqLiteDatabase.rawQuery("select * from orange_user where
            username = ? and password = ?", new String[]{username, password});
        if (cursor.getCount() > 0) {
            return true;
        }
        return false;
    }
}
```

上述代码加粗部分的 validateLogin() 方法，首先获取到登录的账号和密码，然后到 orange 数据库下的 orange_user 表中查询是否存在该用户，如果用户名和密码正确则返回 true，否则返回 false，以此作为是否登录成功的标志。

RegisterActivity 负责的是用户注册的实现，就是往数据库增加记录。当用户输入的信息合法时，就进行插入。大部分代码同之前介绍的一样，在此只给出改变的部分代码，如下所示：

```java
    <!-- 注册验证 -->
    public void validateRegister() {
        String username = usernameEdit.getText().toString();
        String password = passwordEdit.getText().toString();
        String surePassword = surePasswordEdit.getText().toString();
        String city = cityText.getText().toString();
        //判断两次密码是否输入一致
        if (password.equals(surePassword)) {
            //可以再进行其他的验证,如是否符合要求等,符合就进行插入数据库操作
            if (!username.equals("") || !password.equals("")) {
                if (!city.equals("")) {
                    Bundle bundle = new Bundle();
                    bundle.putString("username", username);
                    bundle.putString("password", password);
                    bundle.putString("sex", sexStr);
                    bundle.putString("city", city);
                    //使用安卓自带的 SQLite 数据库存储注册的信息
        OrangeDatabase orangeDatabase = new OrangeDatabase(RegisterActivity.this);
            SQLiteDatabase sqLiteDatabase = orangeDatabase.getWritableDatabase();
```

```java
                    insertData(sqLiteDatabase, bundle);
                    Intent intent = new Intent(RegisterActivity.this, IndexActivity.class);
                    intent.putExtras(bundle);
                    startActivity(intent);
                } else {
        Toast.makeText(RegisterActivity.this, "请选择地址", Toast.LENGTH_SHORT).show();
                }
            } else {
    Toast.makeText(RegisterActivity.this,"账号或密码未填写", Toast.LENGTH_SHORT).show();
            }
        } else {
    Toast.makeText(RegisterActivity.this,"两次密码不一致", Toast.LENGTH_SHORT).show();
        }
    }
    @Override
    public void onCheckedChanged(RadioGroup group, int checkedId) {
        //根据用户选择来改变性别的值
        sexStr = checkedId == R.id.reg_man ? "男" : "女";
    }
    <!-- 插入数据库的值 -->
    private void insertData(SQLiteDatabase sqLiteDatabase, Bundle bundle) {
        ContentValues contentValues = new ContentValues();
        contentValues.put("username", bundle.getString("username"));
        contentValues.put("password", bundle.getString("password"));
        contentValues.put("sex", bundle.getString("sex"));
        contentValues.put("city", bundle.getString("city"));
        sqLiteDatabase.insert("orange_user", null, contentValues);
        sqLiteDatabase.close();
    }
```

从上述代码可以看到,插入数据库操作定义了 insertData()方法,插入的信息有用户名、密码、性别及城市。validateRegister()方法负责对输入数据进行验证,验证通过即可往数据库插入新的数据。生成的数据库如图 6-32 所示。

图 6-32 数据库界面

利用注册过的账号去登录,即可进入到主页中,本节对主页的内容进行调整,方便后续功能的添加和使用。主要用到的知识还是前面已经学习过的。登录成功后进入的主页界面如图 6-33 所示。

图 6-33 "橙一色"App 主页界面

从图 6-33 可以看到,"橙一色"App 主页界面中包含一个底部的导航,导航中含有首页、商品、购物车以及我 4 个功能。单击对应的功能即可在导航的上面展示其对应的界面。关于我的功能在 4.5 节已经实现(见图 4-6),本节主要介绍底部导航的实现和首页信息的展示。首页包含了搜索框、商品分类和模拟的一些热门商品信息。

**1. 底部导航实现**

底部导航采用 Fragment 实现,类似于前面 5.5 节介绍的商品分类功能。首先定义 4 个 Fragment 用于展示不同功能对应的页面。其定义与第 5 章一样,本节就列出 IndexFragment 的内容。具体代码如下:

```
public class IndexFragment extends Fragment implements View.OnClickListener {
    private SearchView searchView;
    private LinearLayout orangeLine, youziLine, juziLine, xiguaLine, liLine, appleLine,
                        lemonLine, mangguoLine;
    private GridView gridView;
    private List<Product> productList;
    private ProductAdapter productAdapter;
    @Override
    public View onCreateView(LayoutInflater inflater, ViewGroup container, Bundle savedInstanceState) {
        View view = LayoutInflater.from(getActivity()).inflate(R.layout.content_index,
                                container, false);
        init(view);
```

```java
        return view;
}
//初始化组件
private void init(View view) {
    searchView = view.findViewById(R.id.searchView);
    searchView.setOnClickListener(this);
    orangeLine = view.findViewById(R.id.chengzi);
    orangeLine.setOnClickListener(this);
    youziLine = view.findViewById(R.id.youzi);
    youziLine.setOnClickListener(this);
    juziLine = view.findViewById(R.id.juzi);
    juziLine.setOnClickListener(this);
    xiguaLine = view.findViewById(R.id.xigua);
    xiguaLine.setOnClickListener(this);
    liLine = view.findViewById(R.id.li);
    liLine.setOnClickListener(this);
    lemonLine = view.findViewById(R.id.lemon);
    lemonLine.setOnClickListener(this);
    mangguoLine = view.findViewById(R.id.mangguo);
    mangguoLine.setOnClickListener(this);
    appleLine = view.findViewById(R.id.apple);
    appleLine.setOnClickListener(this);
    gridView = view.findViewById(R.id.index_famous_gridview);
    initData();
    productAdapter = new ProductAdapter(getActivity(), productList);
    gridView.setAdapter(productAdapter);
}
@Override
public void onClick(View v) {
}
//初始化商品数据
private void initData() {
    productList = new ArrayList< Product >();
    Product product = new Product();
    product.setImageUrlId(R.drawable.juzip);
    product.setProductName("橘子");
    product.setProductPrice(new BigDecimal("9.9"));
    Product product1 = new Product();
    product1.setImageUrlId(R.drawable.orange);
    product1.setProductName("橙子");
    product1.setProductPrice(new BigDecimal("29.9"));
    Product product2 = new Product();
    product2.setImageUrlId(R.drawable.youzip);
    product2.setProductName("柚子");
    product2.setProductPrice(new BigDecimal("19.9"));
    Product product3 = new Product();
    product3.setImageUrlId(R.drawable.xiguap);
    product3.setProductName("西瓜");
```

```
            product3.setProductPrice(new BigDecimal("19.9"));
            Product product4 = new Product();
            product4.setImageUrlId(R.drawable.applep);
            product4.setProductName("苹果");
            product4.setProductPrice(new BigDecimal("49.9"));
            Product product5 = new Product();
            product5.setImageUrlId(R.drawable.lemonp);
            product5.setProductName("柠檬");
            product5.setProductPrice(new BigDecimal("9.9"));
            productList.add(product);
            productList.add(product1);
            productList.add(product2);
            productList.add(product3);
            productList.add(product4);
            productList.add(product5);
    }
}
```

上述代码中 initData()是一个初始化数据的方法,用于模拟数据显示商品的信息。

定义完 Fragment 以后,需要设置底部导航的按钮单击事件,这里需要定义一个首页的 IndexActivity 和其对应的 activity_main.xml 布局文件。其中布局文件的代码如下所示:

```xml
<?xml version = "1.0" encoding = "utf-8"?>
<LinearLayout xmlns:android = "http://schemas.android.com/apk/res/android"
    xmlns:tools = "http://schemas.android.com/tools"
    android:id = "@ + id/activity_main"
    android:layout_width = "match_parent"
    android:layout_height = "match_parent"
    android:orientation = "vertical"
    tools:context = ".activity.IndexActivity">
    <LinearLayout
        android:layout_width = "match_parent"
        android:layout_height = "0dp"
        android:layout_weight = "5">
        <include layout = "@layout/content_user"/>
    </LinearLayout> -->
    <FrameLayout
        android:id = "@ + id/main_content"
        android:layout_width = "match_parent"
        android:layout_height = "0dp"
        android:layout_weight = "1" />
    <LinearLayout
        android:layout_width = "match_parent"
        android:layout_height = "70dp">
        <!-- 底部导航 -->
        <include layout = "@layout/content_nav" />
    </LinearLayout>
</LinearLayout>
```

content_nav.xml 文件是负责设计底部的导航布局,具体代码如下所示:

```xml
<?xml version = "1.0" encoding = "utf-8"?>
<LinearLayout xmlns:android = "http://schemas.android.com/apk/res/android"
    android:layout_width = "match_parent"
    android:layout_height = "match_parent"
    android:orientation = "horizontal">
    <!-- 首页 -->
    <LinearLayout
        android:id = "@ + id/content_index"
        android:layout_width = "0dp"
        android:layout_height = "wrap_content"
        android:layout_weight = "1"
        android:orientation = "vertical">
        <ImageView
            android:layout_width = "wrap_content"
            android:layout_height = "wrap_content"
            android:layout_gravity = "center"
            android:layout_marginTop = "2dp"
            android:src = "@drawable/index" />
        <TextView
            android:layout_width = "wrap_content"
            android:layout_height = "wrap_content"
            android:layout_gravity = "center"
            android:text = "首页"
            android:textColor = "#000"
            android:textSize = "18sp" />
    </LinearLayout>
</LinearLayout>
```

具体的实现效果如图 6-34 所示。

图 6-34 "橙一色"App 主界面

图 6-34 上面的空白区域是一个 FrameLayout 布局,根据选择的功能来展示其对应的界面。"橙一色"App 选择首页作为默认的主界面,所以需要在 IndexActivity 中进行设置。具体代码如下所示:

```java
public class IndexActivity extends Activity implements View.OnClickListener {
    private IndexFragment indexFragment;
    private ProductFragment productFragment;
    private ShoppingCartFragment shoppingCartFragment;
    private PearsonFragment pearsonFragment;
    private LinearLayout indexLine, productLine, shoppingCartLine, pearsonLine;
    @Override
    public void onCreate(@Nullable Bundle savedInstanceState) {
        super.onCreate(savedInstanceState);
        setContentView(R.layout.activity_main);
        init();
        initIndexFragment();
    }
    //组件初始化
    private void init() {
        indexLine = findViewById(R.id.content_index);
        indexLine.setOnClickListener(this);
        productLine = findViewById(R.id.content_product);
        productLine.setOnClickListener(this);
        shoppingCartLine = findViewById(R.id.content_cart);
        shoppingCartLine.setOnClickListener(this);
        pearsonLine = findViewById(R.id.content_pearson);
        pearsonLine.setOnClickListener(this);
    }
    @Override
    public void onClick(View v) {
        switch (v.getId()) {
            case R.id.content_index:
                initIndexFragment();
                break;
            case R.id.content_product:
                initproductFragment();
                break;
            case R.id.content_cart:
                initshoppingCartFragment();
                break;
            case R.id.content_pearson:
                initpearsonFragment();
                break;
        }
    }
    //初始化首页 Fragment
    private void initIndexFragment() {
        //开启事务,Fragment 的控制是由事务来实现的
        FragmentTransaction transaction = getFragmentManager().beginTransaction();
        if (indexFragment == null) {
            indexFragment = new IndexFragment();
        }
```

```
            transaction.replace(R.id.main_content, indexFragment);
            transaction.commit();
    }
    //初始化产品Fragment
    private void initproductFragment() {
        //开启事务,Fragment的控制是由事务来实现的
        FragmentTransaction transaction = getFragmentManager().beginTransaction();
        if (productFragment == null) {
            productFragment = new ProductFragment();
        }
        transaction.replace(R.id.main_content, productFragment);
        transaction.commit();
    }
    //初始化购物车Fragment
    private void initshoppingCartFragment() {
        //开启事务,Fragment的控制是由事务来实现的
        FragmentTransaction transaction = getFragmentManager().beginTransaction();
        if (shoppingCartFragment == null) {
            shoppingCartFragment = new ShoppingCartFragment();
        }
        transaction.replace(R.id.main_content, shoppingCartFragment);
        transaction.commit();
    }
    //初始化个人Fragment
    private void initpearsonFragment() {
        //开启事务,Fragment的控制是由事务来实现的
        FragmentTransaction transaction = getFragmentManager().beginTransaction();
        if (pearsonFragment == null) {
            Intent intent = IndexActivity.this.getIntent();
            Bundle bundle = intent.getExtras();
            pearsonFragment = new PearsonFragment();
            pearsonFragment.setArguments(bundle);
        }
        transaction.replace(R.id.main_content, pearsonFragment);
        transaction.commit();
    }
}
```

可以看到IndexActivity中的代码主要是增加了初始化Fragment的方法,从加粗部分的代码看到,"橙一色"App选择了IndexFragment作为默认的初始化界面,因此当登录或者注册成功以后,即可看到该界面。其他的界面如商品对应的界面、购物车暂时未设计,将在第12章中进行设计补全。

2. 首页信息展示

第一部分完成了底部导航的实现,已经看到了首页具备的信息,IndexFragment的代码已经在上部分进行了介绍,下面主要介绍首页的布局。

首页布局包含搜索框、商品分类和热门商品展示。搜索框使用SearchView组件实现;商品分

类使用线性布局,也可以使用表格布局;热门商品展示使用 GridView 组件,设置 2 列显示。布局代码如下:

```xml
<?xml version = "1.0" encoding = "utf-8"?>
< LinearLayout xmlns:android = "http://schemas.android.com/apk/res/android"
    android:layout_width = "match_parent"
    android:layout_height = "match_parent"
    android:background = "#E8E8E8"
    android:orientation = "vertical">
    < SearchView
        android:id = "@ + id/searchView"
        android:layout_width = "match_parent"
        android:layout_height = "wrap_content"
        android:background = "@drawable/index_menu"
        android:focusable = "false"
        android:iconifiedByDefault = "false"
        android:queryHint = "请输入搜索内容" />
    < LinearLayout
        android:layout_width = "match_parent"
        android:layout_height = "wrap_content"
        android:layout_marginLeft = "1dp"
        android:layout_marginTop = "20dp"
        android:layout_marginRight = "1dp"
        android:background = "@drawable/index_menu"
        android:orientation = "vertical">
        < LinearLayout
            android:layout_width = "match_parent"
            android:layout_height = "wrap_content"
            android:layout_marginTop = "5dp"
            android:orientation = "horizontal">
            < LinearLayout
                android:id = "@ + id/chengzi"
                android:layout_width = "0dp"
                android:layout_height = "wrap_content"
                android:layout_weight = "1"
                android:gravity = "center"
                android:orientation = "vertical">
                < ImageView
                    android:layout_width = "wrap_content"
                    android:layout_height = "wrap_content"
                    android:src = "@drawable/chengzi" />
                < TextView
                    android:layout_width = "wrap_content"
                    android:layout_height = "wrap_content"
                    android:layout_marginTop = "5dp"
                    android:text = "橙子"
                    android:textColor = "#696969"
                    android:textSize = "20sp" />
```

```xml
            </LinearLayout>
            ...
        </LinearLayout>
    <include layout = "@layout/index_famous"/>
</LinearLayout>
```

上述布局文件使用 include 引入了 index_famous 布局文件,用来展示热门商品。具体代码如下所示:

```xml
<?xml version = "1.0" encoding = "utf-8"?>
<LinearLayout xmlns:android = "http://schemas.android.com/apk/res/android"
    android:layout_width = "match_parent"
    android:layout_height = "wrap_content"
    android:layout_marginTop = "15dp"
    android:orientation = "vertical">
    <!-- 热门商品 -->
    <LinearLayout
        android:id = "@+id/index_famous_product"
        android:layout_width = "match_parent"
        android:layout_height = "wrap_content"
        android:layout_marginTop = "15dp"
        android:background = "#FFF"
        android:orientation = "horizontal">
        <TextView
            android:layout_width = "0dp"
            android:layout_height = "50dp"
            android:layout_marginLeft = "5dp"
            android:layout_weight = "1"
            android:drawableLeft = "@drawable/shuiguo"
            android:drawablePadding = "8dp"
            android:gravity = "center_vertical"
            android:text = "热门商品"
            android:textColor = "#000"
            android:textSize = "18sp" />
        <ImageView
            android:layout_width = "25dp"
            android:layout_height = "25dp"
            android:layout_gravity = "center"
            android:src = "@drawable/arrow_right" />
    </LinearLayout>
    <LinearLayout
        android:layout_width = "match_parent"
        android:layout_height = "match_parent"
        android:layout_marginTop = "10dp"
        android:layout_marginLeft = "10dp"
        android:layout_marginRight = "10dp"
        android:orientation = "vertical">
```

```
        <GridView
            android:id = "@ + id/index_famous_gridview"
            android:layout_width = "match_parent"
            android:layout_height = "wrap_content"
            android:background = "#E8E8E8"
            android:verticalSpacing = "10dp"
            android:horizontalSpacing = "10dp"
            android:numColumns = "2" />
    </LinearLayout>
</LinearLayout>
```

通过以上布局，即可完成首页的展示。关于热门商品的数据，在 IndexFragment 中已经定义，是通过一个初始化数据的 initData() 方法来模拟的数据。在 Android Studio 中预览效果如图 6-35 所示。

图 6-35　首页布局预览图

## 6.8　本章小结

本章主要讲解 Android 中的数据存储以及 SQLite 数据库的使用，首先介绍 Android 中常见的几种数据存储方式的异同，讲解 XML 文件序列化与解析、文件存储、SharedPreferences、SQLite 数据库的增加、删除、查找、修改操作，然后讲解 Android 和 Java 对 JSON 文件的解析，最后介绍了"橙一色"App 的数据存储、主界面和底部导航栏的设计。本章所讲解的知识需要熟练掌握，特别是 XML、JSON 文件的使用，这些在 Android 开发中会经常用到。

## 6.9 课后习题

1. 简介几种数据存储方式的各自特点。
2. 使用 SharedPreferences 保存用户登录的信息,并且在用户下次登录时自动填写用户已经保存的信息。
3. 自定义 XNL 文件,并且将内容解析出来。
4. 使用 SQLite 数据库和 ListView 显示一个仓库信息,数据库以及数据通过代码创建和添加。
5. 使用 Java 生成一个 JSON 文件,内容自定。

# ContentProvider 的使用

**学习目标**
- 掌握 ContentProvider 的基本概念。
- 掌握 ContentProvider 的操作。
- 掌握 ContentProvider 的数据共享。
- 掌握 ContentObserver 的使用。

在 Android 开发中,经常需要使用到其他程序中的数据,例如获取用户的手机通讯录、获取发送的验证码。为了实现这种数据的共享使用,Android 系统提供了一个组件 ContentProvider(内容提供者),本章将对 ContentProvider 的使用做详细的讲解。

## 7.1 ContentProvider 简介

Android 官方指出的数据存储方式总共有 5 种,分别是 SharedPreferences、网络存储、文件存储、外部存储、SQLite。但是一般这些存储都只是在单独的一个应用程序之中实现一个数据的共享,有些情况下应用程序需要操作其他应用程序的一些数据,例如获取操作系统里的媒体库、通讯录等,这时就可能通过 ContentProvider 来实现了。

ContentProvider 是 Android 系统的四大组件之一,用于存储和检索数据,是 Android 系统中不同应用程序之间共享数据的接口。它以 Uri 的形式对外提供数据,允许其他应用操作本应用程序的数据,其他应用通过 ContentResolver 提供的 Uri 操作指定的数据。

下面将通过具体的图展示 A 应用程序与 B 应用程序之间的数据共享,如图 7-1 所示。

从图 7-1 可以看出,A 应用程序通过 ContentProvider 将数据暴露出来,供其他应用程序操作。B 应用程序通过 ContentResolver 接口操作暴露的数据,A 应用程序将数据返回到 ContentResolver,然后 ContentResolver 再把数据返回到 B 应用程序。

图 7-1 应用程序共享原理图

## 7.2 操作 ContentProvider

本节主要讲解如何创建和注册 ContentProvider、Uri 参数的基本应用以及使用 ContentProvider 获取用户通讯录。

### 7.2.1 ContentProvider 的创建

开发 ContentProvider 时只需要两步：首先创建一个它的子类，该类需要实现它的抽象方法，如 query()、insert()、update() 和 delete() 等方法；然后在 AndroidManifest.xml 文件中注册 ContentProvider。

上面几个抽象方法的具体的作用如下所示：

- public boolean onCreate()：ContentProvider 创建时调用。
- public int delete()：根据传入的 Uri 删除指定条件下的数据。
- public Uri insert()：根据传入的 Uri 插入数据。
- public Cursor query()：根据传入的 Uri 查询指定的数据。
- public int update()：根据传入的 Uri 更新指定的数据。

为了更好地讲解 ContentProvider，接下来将通过具体的代码讲解 CntentProvider 的创建以及注册。

**1. 创建 ContentProvider**

创建一个类继承 ContentProvider，实现它的抽象方法。具体代码如下所示：

```
public class TestContentProvider extends ContentProvider {
    public boolean onCreate() {
        return false;
    }
    public Cursor query(Uri uri, String[] strings, String s, String[] strings1, String s1) {
        return null;
    }
    public String getType(Uri uri) {
        return null;
    }
    public Uri insert(Uri uri, ContentValues contentValues) {
```

```
        return null;
    }
    public int delete(Uri uri, String s, String[] strings) {
        return 0;
    }
    public int update(Uri uri, ContentValues contentValues, String s, String[] strings) {
        return 0;
    }
}
```

**2. 注册 ContentProvider**

由于 ContentProvider 是 Android 的四大组件之一，所以也需要在 AndroidManifest.xml 文件中注册已经定义的 ContentProvider 子类。具体代码如下所示：

```
< provider
  android:authorities = "com.jxust.cn.chapter7_contentprovider.TestContentProvider "
  android:name = ".TestContentProvider">
</provider >
```

上述就是注册 ContentProvider 的代码，注册时指定了两个属性 android:name 和 android:authorites。android:name 代表的是 ContentProvider 子类的类名，android:authorities 代表的是访问本 provider 的路径。

为了保证数据的安全，有时需要为 Provider 添加权限，这些都可以在清单文件中通过 android:permission 来添加，这里不做具体讲解，用到时可以查一下。

### 7.2.2 Uri 简介

在讲解创建 ContentProvider 时，提到了一个参数 Uri，它代表了数据的操作方法。Uri 由 scheme、authorites、path 3 部分组成，其中 scheme 部分 content://是一个标准的前缀，表明了这个数据被 ContentProvider 管理，它不会修改。authorites 部分是在清单文件注册的 android:authorites 属性值，该值唯一，表明了当前的 ContentProvider。path 部分代表数据，当访问者操作不同数据时，这个值是动态变化的。

以下是 Android 系统提供的一些示例 Uri：

content://media/internal/images——将返回设备上存储的所有图片。

content://contacts/people/——返回设备上的所有联系人信息。

content://contacts/people/45——返回单个结果（联系人信息中 ID 为 45 的联系人记录）。

以上就是 ContentProvider 和 Uri 的简介，接下来将通过获取手机通讯录的案例来具体讲解 ContentProvider 的使用。

### 7.2.3 使用 ContentProvider 获取通讯录

**1. activity_main.xml 文件的编辑**

修改布局文件，采用线性布局的方式，添加一个 ListView 用来显示获取到的姓名和手机号。

视频讲解

activity_main.xml文件的具体代码如下所示：

```xml
<LinearLayout xmlns:android = "http://schemas.android.com/apk/res/android"
    xmlns:tools = "http://schemas.android.com/tools"
    android:layout_width = "match_parent"
    android:layout_height = "match_parent"
    tools:context = "com.jxust.cn.chapter7_getnumber.MainActivity"
    android:orientation = "vertical">
<ListView
    android:id = "@+id/show_people"
    android:layout_width = "match_parent"
    android:layout_height = "wrap_content">
</ListView>
</LinearLayout>
```

**2. MainActivity中的代码**

Android 6.0以后加强了用户权限的设置，首次加载运行App时需要授权同意（Allow）方可读取系统资源，因而在MainActivity中，首先定义了check（）方法进行权限判断，重写了onRequestPermissionsResult（）方法，如果已授权就调用getPhoneName（）方法读取电话联系人信息。getPhoneName（）方法中首先定义获取系统通讯录的Uri，然后获取电话本开始一项的Uri，最后逐行读取，把信息存储到List数组中，利用ArrayAdapter将List数组中的数据显示到ListView中。具体代码如下所示：

```java
public class MainActivity extends Activity {
    @Override
    protected void onCreate(Bundle savedInstanceState) {
        super.onCreate(savedInstanceState);
        setContentView(R.layout.activity_main);
        check();
    }
    private void check() {
        //判断是否有权限
        if (ContextCompat.checkSelfPermission(MainActivity.this,
            Manifest.permission.READ_CONTACTS)!=
            PackageManager.PERMISSION_GRANTED){
            ActivityCompat.requestPermissions(MainActivity.this,
                new String[]{Manifest.permission.READ_CONTACTS},201);
        }else{
            getPhoneName();
        }
    }
    @Override
    public void onRequestPermissionsResult(int requestCode, @NonNull String[] permissions,
            @NonNull int[] grantResults) {
        super.onRequestPermissionsResult(requestCode,permissions, grantResults);
        if (requestCode == 201) {
```

```
            getPhoneName();
        }else{
            return;
        }
    }
    private void getPhoneName() {
        List < String > string;
        //得到 ContentResolver 对象
        ContentResolver cr = getContentResolver();
        Cursor cursor = null;
        try {
        //取得电话本中开始一项的光标
        cursor = cr.query(ContactsContract.Contacts.CONTENT_URI, null, null, null, null);
        }catch (Exception e) {
            Log.i("content_phone", "e = " + e.toString());
        }
        string = new ArrayList < String >();
        //向下移动光标
        while(cursor.moveToNext()) {
          //取得联系人名字
          int nameFieldColumnIndex = cursor.getColumnIndex(
                ContactsContract.PhoneLookup.DISPLAY_NAME);
          String contact = cursor.getString(nameFieldColumnIndex);
          //取得电话号码
          String ContactId =
            cursor.getString(cursor.getColumnIndex(ContactsContract.Contacts._ID));
          Cursor phone =
          cr.query(ContactsContract.CommonDataKinds.Phone.CONTENT_URI, null,
                ContactsContract.CommonDataKinds.Phone.CONTACT_ID + " = " +
                                                    ContactId, null, null);
          while(phone.moveToNext()) {
            String PhoneNumber = phone.getString(
            phone.getColumnIndex(ContactsContract.CommonDataKinds.Phone.NUMBER));
            string.add (contact + ": " + PhoneNumber + "\n");
          }
        }
        cursor.close();
        //获取定义的 ListView 用来显示通讯录信息
        ListView peo_list = findViewById(R.id.show_people);
        peo_list.setAdapter(new ArrayAdapter < String >(MainActivity.this,
                            android.R.layout.simple_list_item_1, string));
    }
}
```

以上就是获取手机通讯录 MainActivity.java 中的详细代码。

### 3. 添加权限

在完成布局文件和 Activity 的代码编写以后,需要在清单文件中添加联系人读写权限。具体代码如下所示：

```
<uses-permission android:name="android.permission.READ_CONTACTS"/>
```

**4. 运行程序**

完成了上述编码以后，运行程序，首次运行会弹出如图 7-2 所示的许可验证提示，单击 ALLOW 后，程序会把系统上的手机通讯录信息读取出来，读取的信息包括联系人姓名和手机号，然后通过 ListView 显示出来，具体的结果如图 7-3 所示。

图 7-2 读取手机通讯录许可

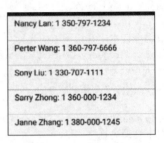
图 7-3 手机通讯录信息

以上就实现了使用系统提供的自带的 Uri 来获取手机通讯录的信息。通过系统自带的这些 Uri 还可以获取媒体信息、短信等。如果想自定义 ContentProvider 类来获取用户信息，可以借助 SQLite 数据库自定义用户数据来实现。

## 7.3 使用 ContentProvider 共享数据

Android 系统应用一般会对外提供 ContentProvider 接口，例如短信、图片以及手机联系人信息等，以便实现应用程序之间的数据共享。应用程序之间的数据共享操作通过 ContentResolver 进行，接下来将详细地介绍 ContentResolver 的相关知识。

### 7.3.1 ContentResolver 简介

在 Android 系统中，应用程序通过 ContentProvider 来暴露自己的数据，然后其他的应用程序通过 ContentResolver 对应用程序暴露的数据进行操作。由于使用 ContentProvider 暴露数据时，提供了相应操作的 Uri，所以在使用 ContentResolver 获取数据的时候，需要指定相应的 Uri，具体代码如下所示：

```
//得到 ContentResolver 对象
ContentResolver cr = getContentResolver();
//取得电话本中开始一项的光标
//ContactsContract.Contacts.CONTENT_URI 手机通讯录的 Uri
Cursor cursor = cr.query(ContactsContract.Contacts.CONTENT_URI, null, null, null, null);
string = new ArrayList<String>();
//向下移动光标
```

```
while(cursor.moveToNext())
{
    ...;
}
cursor.close();
```

在上述代码中,通过 ContentResolver 的 query()方法实现了对应用程序数据的查询,这个方法只适用于查询,不适用于其他的更新、删除等操作。

### 7.3.2 系统短信备份案例

在学习了如何使用 ContentResolver 访问系统应用程序共享的数据之后,下面将给出一个具体的案例——获取系统短信并保存成 XML 文件。具体的步骤如下。

**1. 创建 Android 程序 Chapter7_Sms**

创建完程序以后,首先需要修改布局文件 activity_main.xml,把布局文件修改为线性布局方式,然后添加一个按钮组件,单击按钮以后会读取系统的短信并在/data/data/的项目包下保存一个文件。

activity.xml 文件的具体代码如下所示:

```
<?xml version = "1.0" encoding = "utf-8"?>
<LinearLayout xmlns:android = "http://schemas.android.com/apk/res/android"
    xmlns:tools = "http://schemas.android.com/tools"
    android:layout_width = "match_parent"
    android:layout_height = "match_parent"
    tools:context = "com.jxust.cn.chapter7_sms.MainActivity"
    android:orientation = "vertical">
<Button
    android:id = "@ + id/button1"
    android:layout_width = "match_parent"
    android:layout_height = "wrap_content"
    android:text = "短信备份"/>
</LinearLayout>
```

以上就是程序的布局界面代码,实现的效果如图 7-4 所示。

图 7-4 布局界面

## 2. 编写 SmsInfo 类

因为获取到的短信息较多,需要使用到 List,而且每条短信包含的内容较多,如信息内容、时间、类型、发送地址,此时就需要一个类封装这些信息,这样能够减少代码量,使代码更加容易理解。SmsInfo.java 的具体代码如下所示:

```java
class SmsInfo {
    private String address;                //发送地址
    private long date;                     //发送时间
    private int type;                      //类型
    private String body;                   //内容
    private int id;
    //构造方法
    public SmsInfo(String address, long date, int type, String body) {
        this.address = address;
        this.date = date;
        this.type = type;
        this.body = body;
    }
    public SmsInfo(String address, long date, int type, String body, int id) {
        this.address = address;
        this.date = date;
        this.type = type;
        this.body = body;
        this.id = id;
    }
    public void setAddress(String address) {
        this.address = address;
    }
    public String getAddress() {
        return address;
    }
    public void setDate(long date) {
        this.date = date;
    }
    public long getDate() {
        return date;
    }
    public void setBody(String body) {
        this.body = body;
    }
    public String getBody() {
        return body;
    }
    public void setId(int id) {
        this.id = id;
    }
    public int getId() {
```

```
            return id;
        }
        public void setType(int type) {
            this.type = type;
        }
        public int getType() {
            return type;
        }
}
```

上述就是 SmsInfo JavaBean 对象的具体代码,其中封装了 address、date、type、body 以及 id 属性。

### 3. 创建 XML 文件生成类 Sms_Xml.java

该类负责把获取到的信息保存成一个 mes.xml 文件,存放在/data/data/项目包下,具体代码如下所示:

```
class Sms_Xml {
    //将短信内容 List、文件输出流 FileOutputStream 和 Context 作为构造方法参数
    public static void beifen_sms(List<SmsInfo> list, Context context, FileOutputStream fi_out){
        try{
            XmlSerializer serial = Xml.newSerializer();
            //初始化序列号器,指定 XML 数据写入到哪个文件以及编码
            serial.setOutput(fi_out,"utf-8");
            serial.startDocument("utf-8",true);
            //根节点
            serial.startTag(null,"smss");
            for (SmsInfo info : list){
                //构建父节点
                serial.startTag(null, "sms");
                serial.attribute(null, "id", info.getId() + "");
                //body 部分
                serial.startTag(null, "body");
                serial.text(info.getBody());
                serial.endTag(null, "body");
                //address 部分
                serial.startTag(null, "address");
                serial.text(info.getAddress());
                serial.endTag(null, "address");
                //type 部分
                serial.startTag(null, "type");
                serial.text(info.getType() + "");
                serial.endTag(null, "type");
                //date 部分
                serial.startTag(null, "date");
                serial.text(info.getDate() + "");
                serial.endTag(null, "date");
```

```
                //父节点结束
                serial.endTag(null, "sms");
            }
            serial.endTag(null, "smss");
            serial.endDocument();
            fi_out.close();
        Toast.makeText(context,"短信备份成功",Toast.LENGTH_SHORT).show();
        }catch (Exception e){
            e.printStackTrace();
        Toast.makeText(context,"短信备份失败",Toast.LENGTH_SHORT).show();
        }
    }
}
```

以上就是使用了 XmlSerializer 对象把短信内容以 XML 的形式写入到 mes.xml 文件的具体代码。

如果想要将 mes.xml 保存到 SD 卡中，同样如 6.2.1 节介绍的在"设置"中开启允许访问外部存储。写入 SD 卡的代码如下：

```
File file = new File(Environment.getExternalStorageDirectory(), "mes.xml");
FileOutputStream fi_out = new FileOutputStream(file);
```

采用内部存储或外部存储的区别仅仅是 fi_out 的指向不同，生成 XML 文件的方法相同。

### 4. MainActivity 代码（界面交互类）

MainActivity 类负责实现短信的备份功能。与 7.2.3 节的获取通讯录案例相同，首先要获取访问系统短信的许可，因而在 MainActivity 中，首先定义 check() 方法进行权限判断，重写 onRequestPermissionsResult() 方法，如果已授权就调用 backupSMS() 方法实现 SMS 短信读取和备份。backupSMS() 方法中首先需要初始化按钮组件，然后为其添加监听事件，读取短信内容并保存到 List 中。再将 List 和文件名作为参数传递给 Sms_Xml 类，实现短信备份，具体代码如下所示：

```
public class MainActivity extends Activity {
    private Button button;
    private FileOutputStream fos = null;
    @Override
    protected void onCreate(Bundle savedInstanceState) {
        super.onCreate(savedInstanceState);
        setContentView(R.layout.activity_main);
        check();
    }
    private void check() {
        //判断是否有权限
        if (ContextCompat.checkSelfPermission(MainActivity.this,
            Manifest.permission.READ_SMS)
```

```java
                        != PackageManager.PERMISSION_GRANTED){
            ActivityCompat.requestPermissions(MainActivity.this,
                    new String[]{Manifest.permission.READ_SMS},201);
        } else {
            backupSMS();
        }
    }
    @Override
    public void onRequestPermissionsResult(int requestCode, @NonNull String[] permissions, @NonNull int[] grantResults) {
        super.onRequestPermissionsResult(requestCode, permissions, grantResults);
        if (requestCode == 201) {
            backupSMS();
        } else {
            return;
        }
    }
    private void backupSMS() {
        button = findViewById(R.id.button1);
        button.setOnClickListener(new View.OnClickListener() {
            @Override
            public void onClick(View view) {
                //content://sms 查询所有短信的 Uri
                Uri uri = Uri.parse("content://sms/");
                FileOutputStream fos = null;
                //获取 ContentResolver 对象
                ContentResolver contentResolver = getContentResolver();
                Log.i("get_sms",contentResolver.toString());
                //通过 ContentResolver 对象查询系统短信
                Cursor cursor = contentResolver.query(uri,
                new String[]{"address","date","type","body"},null,null,null);
                List<SmsInfo> list = new ArrayList<SmsInfo>();
                while (cursor.moveToNext()){
                    String address = cursor.getString(0);
                    Log.i("get_sms",address);
                    long date = cursor.getLong(1);
                    int type = cursor.getInt(2);
                    String body = cursor.getString(3);
                    SmsInfo smsInfo = new SmsInfo(address,date,type,body);
                    list.add(smsInfo);
                }
                cursor.close();
                try {
                    fos = openFileOutput("mes.xml",MODE_PRIVATE);
                } catch (FileNotFoundException e) {
                    e.printStackTrace();
                }
                Sms_Xml.beifen_sms(list, MainActivity.this,fos);
```

```
            }
        });
    }
}
```

上述代码中,首先使用了 ContentResolver 读取系统的短信,然后将读取的信息保存为 mes.xml 文件。需要注意的是,使用完 Cursor 之后,一定要关闭,否则会造成内存泄漏。

### 5. 添加权限

该程序需要读取系统的短信信息,所以需要在清单文件中添加权限;如果操作 SD 卡,还需要添加 SD 卡的读取权限。具体代码如下所示:

```
< uses-permission android:name = "android.permission.READ_SMS" />
< uses-permission android:name = "android.permission.WRITE_EXTERNAL_STORAGE" />
```

### 6. 运行程序

上述代码完成以后,运行程序。首次运行会弹出如图 7-5 所示的许可验证提示,单击 ALLOW 后,进入"短信备份"主界面。

单击界面中的"短信备份"按钮,弹出"短信备份成功",如图 7-6 所示。

图 7-5  读取手机短信许可　　　　图 7-6  程序运行界面

在/data/data/项目包下查看生成的 mes.xml 文件。双击可以直接打开文件,查看内容,mes.xml 的具体内容如下所示:

```
<?xml version = '1.0' encoding = 'utf-8' standalone = 'yes' ?>
< smss >
    < sms id = "0">
```

```
            <body>good morning.</body>
            <address>13607976666</address>
            <type>2</type>
            <date>1580354557960</date>
        </sms>
        <sms id="0">
            <body>Hello , Janne</body>
            <address>13800001245</address>
            <type>2</type>
            <date>1580351584319</date>
        </sms>
    </smss>
```

在查看完生成的文件内容后,需要与短信的发送记录进行对比。系统短信的发送记录界面如图 7-7 所示。

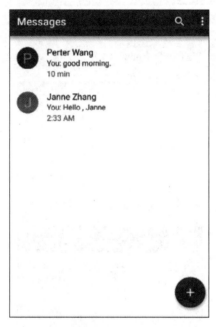

图 7-7　系统短信发送记录

从图 7-7 可以看出,生成的 mes.xml 文件的内容与短信发送的记录内容一致,这就说明读取并保存正确。以上就是使用 ContentResolver 读取系统短信内容的详细使用过程。

### 7.3.3　跨应用数据共享案例

7.2.2 节中介绍了 ContentProvider 的 Uri,列举了 Android 系统提供的关于读取移动设备上的联系人或者 SD 卡上图片的 content://链接,实际应用中,开发者也可以自定义 APP 通过 ContentResolver 实现数据共享。下面给出一个具体的案例,实现两个 App 之间的数据共享。

视频讲解

## 1. 创建 Android 项目 Chapter7_MyCP

创建完项目后，当前默认的 Module 为 App，首先修改 App 的布局文件 activity_main.xml，把布局文件修改为线性布局方式，然后添加一个"写入数据库"的按钮组件，单击按钮后会将数据写入到当前 App 的/data/data/下的数据库中。

activity.xml 文件的具体代码如下所示：

```xml
<?xml version = "1.0" encoding = "utf-8"?>
<LinearLayout
    xmlns:android = "http://schemas.android.com/apk/res/android"
    xmlns:app = "http://schemas.android.com/apk/res-auto"
    xmlns:tools = "http://schemas.android.com/tools"
    android:layout_width = "match_parent"
    android:layout_height = "match_parent"
    android:orientation = "vertical"
    tools:context = ".MainActivity">
    <Button
        android:layout_width = "match_parent"
        android:layout_height = "wrap_content"
        android:layout_marginTop = "10dp"
        android:text = "写入数据库"
        android:textSize = "16dp"
        android:id = "@+id/btnWrite" />
</LinearLayout>
```

以上就是程序的布局界面代码，运行效果如图 7-8 所示。

图 7-8 布局界面

## 2. 编写 MyContentProvider 类

自定义一个 MyContentProvider 类，继承自 ContentProvider。选择 New→Other→ContentProvider，操作界面如图 7-9 所示。

默认类名为 MyContentProvider（可以修改），输入 URI Authorities（通常为项目的包名＋类名缩写），Exported 和 Enabled 默认勾选，为 true，单击 Finish 按钮，创建成功，操作界面如图 7-10 所示。

# 第7章 ContentProvider 的使用

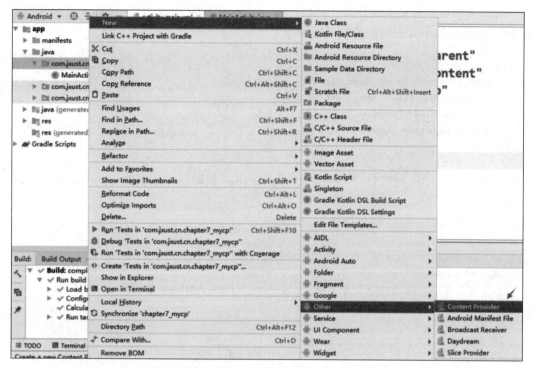

图 7-9 新建 ContentProvider 的子类

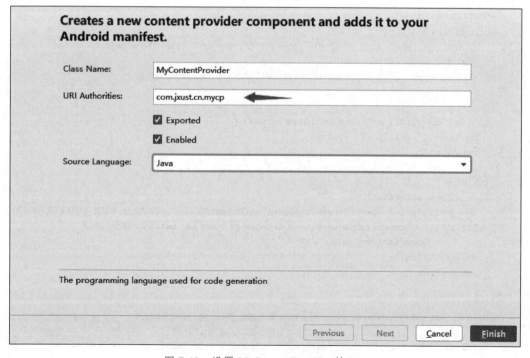

图 7-10 设置 MyContentProvider 的 URI

Provider 为 Android 四大组件之一,创建 MyContentProvider 成功后,会自动添加在清单文件中,AndroidManifest.xml 中的配置如下所示:

```xml
<provider
    android:name=".MyContentProvider"
    android:authorities="com.jxust.cn.mycp"
    android:enabled="true"
    android:exported="true">
</provider>
```

清单中的 authorities 与图 7-10 中的 URI Authorities 相同,enabled 和 exported 为 true,表示允许其他 App 访问。

MyContentProvider.java 的具体代码如下所示:

```java
public class MyContentProvider extends ContentProvider {
    SQLiteDatabase db;
    public final static Uri URI = Uri.parse("content://com.jxust.cn.mycp");
    public MyContentProvider() {
    }
    @Override
    public int delete(Uri uri, String selection, String[] selectionArgs) {
        //Implement this to handle requests to delete one or more rows.
        throw new UnsupportedOperationException("Not yet implemented");
    }
    @Override
    public String getType(Uri uri) {
        //TODO: Implement this to handle requests for the MIME type of the data
        //at the given URI.
        throw new UnsupportedOperationException("Not yet implemented");
    }
    @Override
    public Uri insert(Uri uri, ContentValues values) {
        db.insert("user1","_id",values);
        return null;
    }
    @Override
    public boolean onCreate() {
        db = getContext().openOrCreateDatabase("myContentDB.db3", Context.MODE_PRIVATE, null);
        String sql = "create table user1(_id integer primary key autoincrement," +
                "name text not null)";
        db.execSQL(sql);
        return true;
    }
    @Override
    public Cursor query(Uri uri, String[] projection, String selection,
                        String[] selectionArgs, String sortOrder) {
        Cursor c = db.query("user1",null,null,null,null,null,null);
        return c;
    }
}
```

```
        @Override
        public int update(Uri uri, ContentValues values, String selection,
                          String[] selectionArgs) {
            //TODO: Implement this to handle requests to update one or more rows.
            throw new UnsupportedOperationException("Not yet implemented");
        }
}
```

上述代码中加粗部分是核心代码，Uri 的 content://后面的内容要与清单中的 authorities 的内容一致。在 onCreate()方法中定义 SQLite 数据库 myContentDB.db3，创建表名为 user1 的表。ContentProvider 中提供了 insert()、query()、update()、delete()方法操作数据库。

### 3. 修改 MainActivity 代码

MainActivity 类负责实现数据写入数据库的功能，首先需要初始化按钮组件，然后为其添加监听事件，负责内容存储。具体代码如下所示：

```
public class MainActivity extends AppCompatActivity {
    @Override
    protected void onCreate(Bundle savedInstanceState) {
        super.onCreate(savedInstanceState);
        setContentView(R.layout.activity_main);
                findViewById(R.id.btnWrite).setOnClickListener(new View.OnClickListener() {
            @Override
            public void onClick(View view) {
                write();
            }
            private void write() {
                ContentValues values = new ContentValues();
                values.put("name","JSP");
                getContentResolver().insert(MyContentProvider.URI,values);
                values.put("name","ABC");
                getContentResolver().insert(MyContentProvider.URI,values);
                values.put("name","Excel");
                getContentResolver().insert(MyContentProvider.URI,values);
            }
        });
    }
}
```

上述代码中，利用 getContentResolver()来操作 MyContentProvider.URI 指定的数据库，实现数据的插入。运行程序，单击按钮后，数据存储到数据库中，查询结果如图 7-11 所示。

### 4. 创建读取项目 App2

创建另一个项目读取 Chapter7_MyCP 中 myContentDB.db3 的内容，也可以在同一个项目内创建新的 Module，操作如图 7-12 所示。单击 Module 后，操作与创建新项目的步骤相同，创建成功后的 Module app2 如图 7-13 所示。为了便于区分，将 app2 中的 Activity 命名为 MainActivity2。

图 7-11　ContentProvider 数据库

图 7-12　创建一个新的 Module

图 7-13　Module app2

### 5. activity_main2.xml 代码

app2 中 MainActivity2 对应的布局文件是 activity_main2.xml，定义一个 TextView 用于显示读取的结果。布局代码如下：

```xml
<?xml version = "1.0" encoding = "utf-8"?>
< LinearLayout
    xmlns:android = "http://schemas.android.com/apk/res/android"
    xmlns:app = "http://schemas.android.com/apk/res-auto"
    xmlns:tools = "http://schemas.android.com/tools"
    android:layout_width = "match_parent"
    android:layout_height = "match_parent"
    tools:context = ".MainActivity2">
    < TextView
        android:layout_width = "match_parent"
        android:layout_height = "wrap_content"
        android:layout_marginTop = "10dp"
        android:textSize = "20dp"
        android:textStyle = "bold"
        android:id = "@ + id/tvShow"/>
</LinearLayout >
```

## 6. MainActivity2 代码

MainActivity2 主要实现读取 app 中的数据库内容。其代码如下：

```java
public class MainActivity2 extends AppCompatActivity {
    public final static Uri URI = Uri.parse("content://com.jxust.cn.mycp");
    TextView tvShow;
    @Override
    protected void onCreate(Bundle savedInstanceState) {
        super.onCreate(savedInstanceState);
        setContentView(R.layout.activity_main2);
        tvShow = findViewById(R.id.tvShow);
        Cursor cursor = getContentResolver().query(URI,null,null,null,null);
        cursor.moveToFirst();
        for(int i = 0;i < cursor.getCount();i++){
            String name = cursor.getString(cursor.getColumnIndex("name"));
tvShow.setText(tvShow.getText().toString() + " ======== " + name + " ======== " + "\n");
            cursor.moveToNext();
        }
    }
}
```

上述代码中，Uri 的值要与 app 中 MyContentProvider 的 Uri 相同，说明 app2 通过 getContentResolver()方法读取 app 的 ContentProvider 的数据库。

## 7. 运行程序

上述代码完成以后，运行程序。因为同一个项目中两个 Module，运行时注意选择，操作如图 7-14 所示。app2 的运行结果如图 7-15 所示，读取了 app 中数据库里的数据。

图 7-14　选择 app2 运行

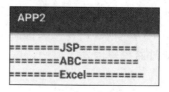

图 7-15　系统短信发送记录

以上就是 ConterntProvider 实现共享数据的详细使用过程。

## 7.4　ContentObserver

### 7.4.1　ContentObserver 简介

ContentObserver(内容观察者)的目的是观察(捕捉)特定 Uri 引起的数据库的变化，继而做一些相应的处理，它类似于数据库技术中的触发器(Trigger)，当 ContentObserver 所观察的 Uri 发生变化时，便会触发 ContentObserver 的 onChange()方法。触发器分为表触发器、行触发器，相应地

ContentObserver 也分为表 ContentObserver、行 ContentObserver,当然这是与它所监听的 Uri MIME Type 有关的。

ContentObserver 的工作原理如图 7-16 所示。

图 7-16 ContentObserver 工作原理

从图 7-16 可以看出,A 应用程序通过 ContentProvider 暴露自己的数据,B 应用程序通过 ContentResolver 操作 A 应用程序的数据,当 A 应用程序的数据发生变化时,A 应用程序调用 notifyChange() 方法向消息中心发送消息,然后 C 应用程序观察到数据变化时,就会触发 ContentObserver 的 onChange() 方法。

接下来讲解一下 ContentObserver 几个常用的方法,具体如下所示:

- public void ContentObserver(Handler handler)——所有 ContentObserver 的派生类都需要调用该构造方法,参数为 handler,Handler 对象。参数 handler 是 Handler 类的对象,可以是主线程 Handler(这时候可以更新 UI),也可以是任何 Handler 对象。
- void onChange(boolean selfChange)——观察到的 Uri 发生变化时,回调该方法去处理。所有 ContentObserver 的派生类都需要重载该方法去处理数据。

接下来将通过监控短信发送案例讲解如何注册内容观察者、自定义观察者以及当数据变化时怎么处理。

### 7.4.2 监控短信发送案例

7.4.1 节学习了 ContentResolver 的基本知识以及常用的方法,本节将通过具体的例子来讲解如何使用。具体的操作步骤如下。

**1. 创建 Chapter7_SmsListener 程序**

本案例通过监听 Uri 为 content://sms 的数据改变即可监听到用户信息的数据改变,并且通过监听器的 onChange(Boolean selfChange) 方法查询 Uri 为 content://sms/outbox 的数据,获取用户正在发送的短信(用户正在发送的短信是保存在发件箱内)。

修改 activity_main.xml 文件代码,采用相对布局的方式,放置一个 TextView 组件,用来显示发送消息的内容。具体代码如下所示:

# 第7章 ContentProvider 的使用

```xml
<?xml version = "1.0" encoding = "utf-8"?>
< RelativeLayout xmlns:android = "http://schemas.android.com/apk/res/android"
    xmlns:tools = "http://schemas.android.com/tools"
    android:layout_width = "match_parent"
    android:layout_height = "match_parent"
    tools:context = "com.jxust.cn.chapter7_smslistener.MainActivity">
    < TextView
        android:layout_width = "wrap_content"
        android:layout_height = "wrap_content"
        android:text = "显示发送消息的内容"
        android:layout_centerHorizontal = "true"
        android:layout_centerVertical = "true"
        android:textSize = "17sp"/>
</RelativeLayout >
```

上述代码定义了一个居中显示的 TextView 组件,用来显示发送消息的内容。具体的界面如图 7-17 所示。

图 7-17 布局界面

### 2. MainActivity 代码

在完成了布局界面的设计以后,接下来需要在 MainActivity 中监听发送的消息,并且把消息显示出来。具体代码如下所示:

```java
public class MainActivity extends Activity {
    private TextView mes_text;
    @Override
    protected void onCreate(Bundle savedInstanceState) {
        super.onCreate(savedInstanceState);
        setContentView(R.layout.activity_main);
        check();
    }
    private void check() {
```

```java
        //判断是否有权限
        if (ContextCompat.checkSelfPermission(MainActivity.this,
                Manifest.permission.READ_SMS)
                    != PackageManager.PERMISSION_GRANTED) {
            ActivityCompat.requestPermissions(MainActivity.this,
                new String[]{Manifest.permission.READ_SMS},201);
        }else{
            mes_listener();
        }
    }
    @Override
    public void onRequestPermissionsResult(int requestCode, @NonNull String[]
                            permissions, @NonNull int[] grantResults) {
        super.onRequestPermissionsResult(requestCode, permissions, grantResults);
        Log.i("SMS_Listener","requestCode = " + requestCode);
        if(requestCode == 201){
            mes_listener();
        }else{
            return;
        }
    }
    private void mes_listener() {
        mes_text = findViewById(R.id.show_mes);
        //为content://sms 的数据改变注册监听器
        ContentResolver contentResolver = getContentResolver();
        Log.i("SMS_Listener","contentResolver = " + contentResolver);
        Uri uri = Uri.parse("content://sms/");
        contentResolver.registerContentObserver(uri,true,new SmsObsever(new Handler()));
    }
    //自定义的ContentObserver监听器类
    private class SmsObsever extends ContentObserver {
        public SmsObsever(Handler handler) {
            super(handler);
            Log.i("SMS_Listener","SmsObsever");
        }
        @Override
        public void onChange(boolean selfChange) {
            Log.i("SMS_Listener","onChange");
            //查询发件箱中的短信
            Cursor cursor = getContentResolver().query(
                        Uri.parse("content://sms/outbox "),null,null,null,null);
            Log.i("SMS_Listener","cursor = " + cursor);
            String message = "";
            //遍历查询的结果集
            while(cursor.moveToNext()){
                String address = cursor.getString(cursor.getColumnIndex("address"));
                String body = cursor.getString(cursor.getColumnIndex("body"));
                String time = cursor.getString(cursor.getColumnIndex("date"));
```

```
            message = message + "收件人:" + address + "\n内容:" + body + "\n发送时间:" + time + "\n";
        }
        mes_text.setText(message);
        cursor.close();
    }
}
```

上述代码首次运行要获得授权许可,mes_listener()方法首先监听 Uri 为 content://sms/的数据改变,从而监听用户信息数据的改变;然后监听 Uri 为 content://sms/outbox 的全部数据,从而查询用户刚发送的短信。

### 3. 添加权限

因为需要读取用户的短信数据,所以需要在清单文件中添加权限。具体代码如下所示:

```
<uses-permission android:name="android.permission.READ_SMS" />
```

### 4. 运行程序

运行该程序,首次运行时要授权允许访问 SMS,如图 7-18 所示,单击 ALLOW,然后跳转到应用程序。在不关闭程序的情况下(按 Home 键返回到桌面)打开系统自带的短信发送程序,发送一条消息,如图 7-19 所示。该程序后台检测到的发送信息的内容如图 7-20 所示。

图 7-18 授权许可

图 7-19 发送信息界面

图 7-20 程序运行结果

以上就是监听用户发送短信的具体操作,这个案例采用 Activity 来实现并不合适,因为用户需要先打开该应用程序,然后在保持该 Activity 不关闭的情况下去发送短信,这样才能监听到发送的短信。这不符合用户的操作,在第 8 章中会讲解如何利用 Android 中的 Service 组件来实现以后台进程的方式监听用户发送的信息。

## 7.5 本章小结

本章主要讲解 Android 系统中 ContentProvider 组件的功能和用法,首先对 ContentProvider 进行简单的介绍,然后讲解如何创建 ContentProvider 以及如何使用 ContentProvider 访问其他应用程序的数据,最后讲解 ContentObserver,通过 ContentObserver 观察数据的变化。

至此,Android 中的四大组件已经讲了 Activity 和本章的 ContentProvider,还有两个组件,分别为 Service、BroadcastReceiver,在接下来的两章中将进行讲解。已经学过的两大组件需要熟练掌握,在实际的项目开发中,它们有很大的用处。

## 7.6 课后习题

1. 简述 ContentProvider 的工作原理。
2. 简述 ContentProvider、ContentResolver、ContentObserver 之间的关系。
3. 使用 ContentObserver 监听用户接收的短信数据,使用 TextView 显示监听到的数据。
4. 自定义联系人数据库,使用 ContentProvider 将联系人的信息显示在 ListView 组件上。

# Service 和广播的使用

**学习目标**
- 掌握 Service 的创建、配置。
- 掌握 Service 的两种启动方式以及停止方式。
- 掌握 Service 的生命周期。
- 掌握 Service 的通信。
- 掌握广播(Broadcast)的使用。

Service 是 Android 四大组件中与 Activity 最相似的组件,它们都代表可执行的程序,都有自己的生命周期,不同的是 Service 一直在后台运行,它没有界面,而 Activity 有自己的运行界面。因此 Service 经常用于处理一些耗时的程序,例如网络传输、视频播放等。

广播就是一种广泛运用在应用程序之间传输消息的机制,而广播接收器(BroadcastReceiver)是 Android 应用程序的四大组件之一,对发送出来的广播进行接收并响应的一类组件。接下来将详细介绍 Service 和广播的使用。

## 8.1 Service 简介

Service 组件是可执行的程序,它能够长期在后台运行且不提供用户界面,它也有自己的生命周期。创建、配置 Service 与创建、配置 Activity 的过程基本相似,只需要继承 Service 类,接下来将详细介绍 Service 的开发。

### 8.1.1 Service 的创建和配置

前面学过 Activity 的创建与配置,首先要创建 Activity 子类,然后在 AndroidManifest.xml 文件中配置 Activity。开发 Service 也需要两步:首先创建 Service 子类,然后在清单文件中配置。Service 与 Activity 都是从 Context 派生出来的,因此都可调用 Context 中定义的如

getResources()、getContentResolver()等方法。

### 1. Service 的创建

创建一个 Test_Service 类继承 Service，此时该类会自动实现 onBind()方法。Test_Service 类的具体代码如下所示：

```java
public class Test_Service extends Service {
    @Override
    public IBinder onBind(Intent intent) {
        return null;
    }
}
```

上述代码中创建了一个 Test_Service 类继承自 Service，在该类中实现了一个 onBind()方法，关于该方法会在后面进行详细讲解。

### 2. Service 的配置

由于 Service 是 Android 的四大组件之一，因此需要在清单文件中注册。注册的具体代码如下所示：

```xml
<application
    ...
    <service android:name=".Test_Service"></service>
</application>
```

以上就是 Service 组件的创建与配置，需要注意的是创建完成以后，一定要在清单文件中配置，否则服务是无效的。

## 8.1.2 Service 的启动与停止

当 Service 创建与配置完成以后，接下来就可以在程序中运行该 Service 了。在 Android 系统中运行 Service 有如下两种方式：

- 通过 Context 的 startService()方法。通过该方法启动 Service，访问者与 Service 之间没有关联，即使访问者退出，Service 也仍然在运行。
- 通过 Context 的 bindService()方法。这种方式启动的 Service，访问者与 Service 绑定在一起，访问者退出，Service 也就终止了。

接下来将详细讲解这两种启动方式。

### 1. startService 方式启动服务

学习 startService 方式启动服务，首先要学会使用 startService()方法开启服务和使用 stopService()关闭服务。具体代码如下所示：

```java
Intent intent = new Intent(this, Test_Service.class);
startService(intent);              //开启服务
stopService(intent);               //关闭服务
```

以上就是开启 Test_Service 服务的方法,接下来将通过具体的例子讲解 startService 方式启动服务的过程。

1)创建 Chapter8_start_Service 项目

首先创建一个项目,然后修改 activity_main.xml 布局文件,采用线性布局的方式,放置两个按钮组件,分别用来启动和关闭服务。代码如下所示:

```xml
<?xml version = "1.0" encoding = "utf-8"?>
< LinearLayout xmlns:android = "http://schemas.android.com/apk/res/android"
    xmlns:tools = "http://schemas.android.com/tools"
    android:layout_width = "match_parent"
    android:layout_height = "match_parent"
    tools:context = "com.jxust.cn.chapter8_start_service.MainActivity"
    android:orientation = "vertical">
    < Button
        android:id = "@ + id/start"
        android:layout_width = "match_parent"
        android:layout_height = "wrap_content"
        android:text = "启动服务" />
    < Button
        android:id = "@ + id/stop"
        android:layout_width = "match_parent"
        android:layout_height = "wrap_content"
        android:text = "停止服务" />
</LinearLayout >
```

上述布局代码实现的界面如图 8-1 所示。

图 8-1　程序界面

2)创建 Service 子类

创建一个 Test_Service 类继承自 Service,重写生命周期中的 onCreate()、onStartCommand()和 onDestroy()方法,然后通过输出 Log 信息,观察服务的执行过程。具体代码如下所示:

```java
public class Test_Service extends Service {
    @Override
```

```java
public IBinder onBind(Intent intent) {
    return null;
}
@Override
public void onCreate() {
    super.onCreate();
    Log.v("test_Service", "onCreate()");
}
@Override
public int onStartCommand(Intent intent, int flags, int startId) {
    Log.v("Test_Service", "onStartCommand()");
    return super.onStartCommand(intent, flags, startId);
}
public void onDestroy() {
    Log.v("Test_Service", "onDestroy()");
    super.onDestroy();
}
}
```

3）配置 Service

清单文件中配置的具体代码如下所示：

```xml
<service android:name=".Test_Service"></service>
```

4）编写 MainActivity 代码

MainActivity 主要负责实现按钮的单击事件，分别是启动服务和关闭服务。具体代码如下所示：

```java
public class MainActivity extends Activity implements View.OnClickListener {
    private Button start_btn, stop_btn;
    @Override
    protected void onCreate(Bundle savedInstanceState) {
        super.onCreate(savedInstanceState);
        setContentView(R.layout.activity_main);
        start_btn = findViewById(R.id.start);
        stop_btn = findViewById(R.id.stop);
        start_btn.setOnClickListener(this);
        stop_btn.setOnClickListener(this);
    }
    @Override
    public void onClick(View view) {
        switch (view.getId()){
            //启动服务
            case R.id.start:
                Intent intent = new Intent(this, Test_Service.class);
                startService(intent);
                break;
```

```
            //关闭服务
            case R.id.stop:
                Intent intent11 = new Intent(this, Test_Service.class);
                stopService(intent11);
                break;
            default:
                break;
        }
    }
}
```

5）运行程序

运行程序，单击界面上的"启动服务"按钮，出现的 Log 信息如图 8-2 所示。

```
2020-01-31 16:16:13.361 21640-21640/com.jxust.cn.chapter8_start_service V/Test_Service: onCreate()
2020-01-31 16:16:13.362 21640-21640/com.jxust.cn.chapter8_start_service V/Test_Service: onStartCommand()
```

图 8-2 "启动服务"结果

从 Log 信息可以看出，服务创建时首先执行的是 onCreate()方法，当服务启动时执行的是 onStartCommand()方法。需要注意的是，onCreate()方法只在服务创建时执行，而 onStartCommand()方法则在每次启动服务时调用。

单击"停止服务"按钮，打印的 Log 信息如图 8-3 所示。

```
2020-01-31 16:16:13.361 21640-21640/com.jxust.cn.chapter8_start_service V/Test_Service: onCreate()
2020-01-31 16:16:13.362 21640-21640/com.jxust.cn.chapter8_start_service V/Test_Service: onStartCommand()
2020-01-31 16:24:15.849 21640-21640/com.jxust.cn.chapter8_start_service V/Test_Service: onDestroy()
```

图 8-3 "停止服务"结果

从图 8-3 所示的 Log 信息可以看出，当单击"停止服务"按钮以后，服务会执行 onDestroy()方法销毁。以上就是使用 startService 方式启动服务的方法，如果不调用 stopService()方法，服务会在后台一直运行，除非用户强制停止服务。

**2. bindService 方式启动服务**

当程序使用 startService()和 stopService()启动和关闭服务时，服务与调用者之间基本不存在太多的关联，因此 Service 无法与访问者之间进行数据交换和通信等。

如果 Service 和访问者之间需要进行通信和数据交换时，则应该使用 bindService()和 unbindService()方法启动、关闭 Service。

Context 的 bindService()方法的完整形式如下所示：

```
bindService(Intent service, ServiceConnection conn, int flags)
```

该方法的 3 个参数的说明如下：

service——该参数通过 Intent 指定要启动的 Service。

conn——该参数是一个 ServiceConnection 对象，用于监听访问者与 Service 之间的连接情况。当访问者与 Service 连接成功时将调用 ServiceConnection 对象的 onServiceConnected(ComponentName

name,IBinder service)方法,当 Service 与访问者之间断开连接时将调用 ServiceConnection 对象的 onServiceDisconnected(ComponentName name)方法。

flags——指定绑定时是否自动创建 Service(如果 Service 还没创建)。该参数可指定为 0(不自动创建)或 BIND_AUTO_CREATE(自动创建)。

实际的开发中通常会采用继承 Binder(IBinder 的实现类)的方式实现自定义的 IBinder 对象。接下来通过一个具体的例子讲解使用 bindService 方式启动和关闭服务的执行过程。

1) 创建 Chapter8_bind_Service 项目

首先修改布局界面的代码,采用线性布局的方式放置 3 个按钮。activity_main.xml 的具体代码如下所示:

```xml
<?xml version = "1.0" encoding = "utf-8"?>
<LinearLayout xmlns:android = "http://schemas.android.com/apk/res/android"
    xmlns:tools = "http://schemas.android.com/tools"
    android:layout_width = "match_parent"
    android:layout_height = "match_parent"
    tools:context = "com.jxust.cn.chapter8_bind_service.MainActivity"
    android:orientation = "vertical">
    <Button
        android:id = "@ + id/button1"
        android:layout_width = "match_parent"
        android:layout_height = "wrap_content"
        android:text = "绑定服务" />
    <Button
        android:id = "@ + id/button2"
        android:layout_width = "match_parent"
        android:layout_height = "wrap_content"
        android:text = "调用服务的方法" />
    <Button
        android:id = "@ + id/button3"
        android:layout_width = "match_parent"
        android:layout_height = "wrap_content"
        android:text = "解除绑定" />
</LinearLayout>
```

上述代码中定义了 3 个按钮,分别对应着绑定服务、调用服务的方法和解除绑定的事件。

上述程序实现的用户界面如图 8-4 所示。

图 8-4　程序界面

2）创建 Service 类

创建一个 Test_bindService 类继承 Service，该类中实现了绑定服务生命周期中的 3 个方法以及自定义的一个 custom_method()方法。该类的具体实现代码如下：

```java
public class Test_bindService extends Service {
    //创建服务的代理,调用服务中的方法
    class MyBinder extends Binder{
        public void test(){
            custom_method();
        }
    }
    //自定义方法
    private void custom_method() {
        Log.v("Test_bindService", "自定义的方法 custom_metod()");
    }
    @Override
    public void onCreate() {
        Log.v("Test_bindService", "创建服务 onCreate()");
        super.onCreate();
    }
    @Override
    public IBinder onBind(Intent intent) {
        Log.v("Test_bindService", "绑定服务 onBind()");
        return new MyBinder();
    }
    @Override
    public boolean onUnbind(Intent intent) {
        Log.v("Test_bindService", "解除绑定 onUnbind()");
        return super.onUnbind(intent);
    }
}
```

3）配置 Service

在清单文件中配置 Service 的代码如下所示：

```xml
<service android:name=".Test_bindService"></service>
```

4）编写 MainActivity 代码

MainActivity 主要负责实现按钮的监听功能，分别对应着绑定服务、调用服务的方法、解除绑定。具体代码如下所示：

```java
public class MainActivity extends Activity implements View.OnClickListener {
    private Button btn1, btn2, btn3;
    private test_bindService.MyBinder myBinder;
    private MyConn conn;
    @Override
    protected void onCreate(Bundle savedInstanceState) {
```

```java
        super.onCreate(savedInstanceState);
        setContentView(R.layout.activity_main);
        btn1 = findViewById(R.id.button1);
        btn2 = findViewById(R.id.button2);
        btn3 = findViewById(R.id.button3);
        btn1.setOnClickListener(this);
        btn2.setOnClickListener(this);
        btn3.setOnClickListener(this);
    }
    @Override
    public void onClick(View view) {
        switch (view.getId()){
            //绑定服务
            case R.id.button1:
                if(conn == null){
                    conn = new MyConn();
                }
                Intent intent = new Intent(this, test_bindService.class);
                bindService(intent, conn, BIND_AUTO_CREATE);
                break;
            //调用服务中的方法
            case R.id.button2:
                myBinder.test();
                break;
            //解除绑定
            case R.id.button3:
                if(conn!= null){
                    unbindService(conn);
                    conn = null;
                }
                break;
            default:
                break;
        }
    }
    //创建MyConn类,用于实现连接服务
    private class MyConn implements ServiceConnection{
        //成功绑定到服务时调用的方法
        @Override
        public void onServiceConnected(ComponentName componentName, IBinder iBinder) {
            Log.v("Test_bindService ", "服务绑定成功");
            myBinder = (test_bindService.MyBinder)iBinder;
        }
        @Override
        public void onServiceDisconnected(ComponentName componentName) {
        }
    }
}
```

5) 运行程序

运行程序,单击界面上的"绑定服务"按钮,打印的Log信息如图8-5所示。

```
2020-01-31 16:40:31.292 22339-22339/com.jxust.cn.chapter8_bind_service V/Test_bindService: 创建服务onCreate()
2020-01-31 16:40:31.292 22339-22339/com.jxust.cn.chapter8_bind_service V/Test_bindService: 绑定服务onBind()
2020-01-31 16:40:31.340 22339-22339/com.jxust.cn.chapter8_bind_service V/Test_bindService: 服务绑定成功
```

图 8-5 "绑定服务"结果

从图 8-5 中可以看出,服务绑定成功了,在服务绑定时会首先调用 onCreate()方法,然后调用 onBind()方法。接下来单击"调用服务的方法"按钮,打印的 Log 信息如图 8-6 所示。

```
2020-01-31 16:40:31.292 22339-22339/com.jxust.cn.chapter8_bind_service V/Test_bindService: 创建服务onCreate()
2020-01-31 16:40:31.292 22339-22339/com.jxust.cn.chapter8_bind_service V/Test_bindService: 绑定服务onBind()
2020-01-31 16:40:31.340 22339-22339/com.jxust.cn.chapter8_bind_service V/Test_bindService: 服务绑定成功
2020-01-31 16:44:43.363 22339-22339/com.jxust.cn.chapter8_bind_service V/Test_bindService: 自定义的方法custom_metod()
```

图 8-6 "调用服务的方法"结果

接下来单击"解除绑定"按钮,打印的 Log 信息如图 8-7 所示。

```
2020-01-31 16:40:31.292 22339-22339/com.jxust.cn.chapter8_bind_service V/Test_bindService: 创建服务onCreate()
2020-01-31 16:40:31.292 22339-22339/com.jxust.cn.chapter8_bind_service V/Test_bindService: 绑定服务onBind()
2020-01-31 16:40:31.340 22339-22339/com.jxust.cn.chapter8_bind_service V/Test_bindService: 服务绑定成功
2020-01-31 16:44:43.363 22339-22339/com.jxust.cn.chapter8_bind_service V/Test_bindService: 自定义的方法custom_metod()
2020-01-31 16:47:39.083 22339-22339/com.jxust.cn.chapter8_bind_service V/Test_bindService: 解除绑定onUnbind()
```

图 8-7 "解除绑定"结果

以上就是使用 bindService 方式启动 Service 的具体操作过程,通过这种形式,Service 可以与访问者之间进行通信与数据交换等。

## 8.2 Service 的生命周期

在 8.1 节学习 Service 的启动与停止时,已经简单了解了 Service 生命周期的几个方法,如 onCreate()、onStartCommand()、onDestroy()方法。接下来将详细介绍在两种不同启动方式下的 Service 的生命周期。

视频讲解

**1. startService 方式启动服务的生命周期**

当使用 startService 方式启动服务时,服务会先执行 onCreate()方法,接着执行 onStartCommand() 方法,此时服务处于运行状态,直到自身调用 stopSelf()方法或者访问者调用 stopService()方法时服务停止,最终被系统销毁。这种方式开启的服务会长期在后台运行,与访问者的状态没有关系。

**2. bindService 方式启动服务的生命周期**

当其他组件调用 bindService()方法时,服务首先被创建,接着访问者通过 IBinder 接口与服务通信。访问者通过 unbindService()方法关闭连接,当多个访问者绑定在一个服务上时,只有它们都解除绑定时,服务才会被直接销毁。这种方式启动的服务与访问者有关,访问者退出时,服务就会被销毁。

**3. 服务生命周期的方法介绍**

onCreate():第一次创建服务时执行的方法。

onDestroy()：服务被销毁时执行的方法。
onStartCommand()：访问者通过 startService(Intent service)启动服务时执行的方法。
onBind()：使用 bindService()方式启动服务调用的方法。
onUnbind()：解除绑定时调用的方法。

上述这些方法都是 Service 生命周期的重要回调方法，通过该方法可以看出服务从启动到停止所经历的过程。

为了更清楚地看到服务两种启动方式的生命周期，接下来将通过一张对比图来说明，如图 8-8 所示。

图 8-8　Service 的生命周期

以上就是两种启动方式的生命周期，和 Activity 的生命周期类似。关于 Service 的基本应用需要熟练掌握，后续章节会讲解如何利用系统的服务，如电话、短信、闹钟等。

## 8.3　Service 通信

### 8.3.1　本地服务和远程服务通信

Android 系统中，服务的通信方式有两种：本地服务通信和远程服务通信。本地服务通信是指应用程序内部的通信,远程服务通信是指两个应用程序之间的通信。无论使用哪一种通信,必须要以绑定的方式开启服务。接下来将详细介绍这两种服务。

1. 本地服务通信

使用服务进行本地通信时,首先需要开发一个 Service 类,该类会提供一个 IBinder onBind(Intent intent)方法,onBind()方法返回的 IBinder 对象会作为参数传递给 ServiceConnection 类中 onServiceConnected(Component name, IBinder service)方法,这样访问者即可通过 IBinder 对象与 Service 进行通信。

接下来将通过一张图来说明如何使用 IBinder 对象进行本地服务通信,如图 8-9 所示。

图 8-9 本地服务通信原理

从图 8-9 中可以看出,本地服务通信实质上是使用了 IBinder 对象,在 ServiceConnection 类中得到 IBinder 对象,就可以获取 Service 中定义的方法。

2. 远程服务通信

远程服务通信主要是为了实现不同应用程序之间的通信,远程服务通信是通过 AIDL(Android Interface Definition Language)实现的,它是一种接口定义语言。开发人员定义的 AIDL 接口只是定义了进程之间的通信接口,服务端、客户端都需要使用 Android SDK 安装目录下的 platform-tools 子目录下的 aidl.exe 工具为接口提供实现。如果开发人员使用 ADT 工具进行开发,那么 ADT 工具会自动实现 AIDL 接口。

定义 AIDL 接口的具体示范代码如下所示:

```
interface IService{
    String getNmae();
    int getAge();
}
```

定义 AIDL 接口时,不需要添加类型修饰符,例如 public、private 等,如果添加则是不正确的。

定义好 AIDL 接口之后,接着创建 Service 类的子类。该 Service 的 onBind()方法返回的 IBinder 对象应该是 ADT 所生成的 IService.Stub 的子类。具体代码如下所示:

```
public class MyService extends Service{
    //继承 IService.stub
    private class IServiceBinder extends Stub {
        public String getName()throws RemoteException{
            return "王思";
        }
        public int getAge()throws RemoteException{
```

```
            return 12;
        }
    }
    @Override
    public IBinder onBind(Intent intent) {
        return new IServiceBinder();
    }
    @Override
    public void onCreate() {
        super.onCreate();
    }
}
```

使用 AIDL 的交互过程是 client <--> proxy <--> stub <--> service，stub 和 proxy 是为了方便客户端/服务器端交互而生成出来的代码，这样客户端/服务器的代码就会比较干净，不会嵌入很多很难懂的与业务无关的代码。

### 8.3.2 本地服务通信实例

视频讲解

8.3.1 节介绍了本地服务通信与远程服务通信，接下来将通过一个如何在 Activity 中绑定本地 Service、传递参数到 Service，并获取 Service 返回数据的案例来讲解本地服务通信的使用方式。具体步骤如下。

**1. 创建 Chapter8_Service_Test 项目**

首先设计用户交互界面，修改 activity_main.xml 布局文件的代码，采用线性布局的方式放置 1 个 TextView、1 个 EditText 和 5 个按钮，TextView 用于显示结果，EditText 用于输入数据，5 个按钮分别是"启动服务""停止服务""绑定服务""解除绑定服务""同步服务数据"。具体代码如下所示：

```xml
<?xml version = "1.0" encoding = "utf-8"?>
<LinearLayout xmlns:android = "http://schemas.android.com/apk/res/android"
    xmlns:app = "http://schemas.android.com/apk/res-auto"
    xmlns:tools = "http://schemas.android.com/tools"
    android:layout_width = "match_parent"
    android:layout_height = "match_parent"
    android:orientation = "vertical"
    tools:context = ".MainActivity">
    <TextView
        android:layout_width = "match_parent"
        android:layout_height = "wrap_content"
        android:layout_marginTop = "6dp"
        android:textSize = "18dp"
        android:textStyle = "bold"
        android:id = "@+id/tvOut"/>
    <EditText
        android:layout_width = "match_parent"
        android:layout_height = "wrap_content"
        android:text = "input message"
```

```xml
        android:textSize = "18dp"
        android:textStyle = "bold"
        android:id = "@ + id/etInput"
        />
    <Button
        android:layout_width = "match_parent"
        android:layout_height = "wrap_content"
        android:layout_marginTop = "6dp"
        android:text = "启动服务"
        android:id = "@ + id/btnStartService"/>
    <Button
        android:layout_width = "match_parent"
        android:layout_height = "wrap_content"
        android:layout_marginTop = "6dp"
        android:text = "停止服务"
        android:id = "@ + id/btnStopService"/>
    <Button
        android:layout_width = "match_parent"
        android:layout_height = "wrap_content"
        android:layout_marginTop = "6dp"
        android:text = "绑定服务"
        android:id = "@ + id/btnBindService"/>
    <Button
        android:layout_width = "match_parent"
        android:layout_height = "wrap_content"
        android:layout_marginTop = "6dp"
        android:text = "解除绑定服务"
        android:id = "@ + id/btnUnBindService"/>
    <Button
        android:layout_width = "match_parent"
        android:layout_height = "wrap_content"
        android:layout_marginTop = "6dp"
        android:text = "同步服务数据"
        android:id = "@ + id/btnSync"/>
</LinearLayout>
```

以上代码实现的用户界面如图8-10所示。

图 8-10　用户界面

## 2. 创建 Service 的子类

创建一个 Service 的子类 MyService，该类负责实现 onCreate()方法、onStartCommand()方法、onBind()方法、onUnbind()方法、onDestroy()方法。另外需要创建一个内部类继承自 Binder，负责实现通信。MyService 的具体代码如下所示：

```java
public class MyService extends Service{
    boolean running = false;                    //定义启动服务和停止服务的标志
    String str = "来自服务器的信息...";           //初始默认的显示信息
    public MyService() {
    }
    @Override
    public IBinder onBind(Intent intent) {
        //TODO: Return the communication channel to the service.
        Log.i("MyService", "onBind()");
        return new MyBinder();                  //匿名类形式返回 MyBinder 类的对象
    }
    //自定义内部类 MyBinder 类继承 Binder 类
    class MyBinder extends Binder{
        void setData(String s){
            MyService.this.str = s;             //修改 MyService 中的值
        }
        MyService getService() {
            return MyService.this;
        }
    }
    @Override
    public void onCreate() {
        super.onCreate();
        Log.i("MyService", " onCreate ()");
        //启动一个新的线程
        new Thread(){
            @Override
            public void run() {
                super.run();
                running = true;
                int i = 1;
                while(running) {
                    try {
                        String s1 = i + ":" + str;
                        Log.i("MyService", s1);
                        i++;
                        if(callback!= null){
                            //调用接口中的方法获取数据
                            callback.dataChanged(s1);
                        }
                        sleep(1000);
                    } catch (InterruptedException e) {
```

```
                    e.printStackTrace();
                }
            }
        }
    }.start();
}
@Override
public int onStartCommand(Intent intent, int flags, int startId) {
    Log.i("MyService", " onStartCommand ()");
    str = intent.getStringExtra("data");              //intent 传递参数
    return super.onStartCommand(intent, flags, startId);
}
@Override
public void onDestroy() {
    super.onDestroy();
    Log.i("MyService", " onDestroy ()");
    running = false;
}
@Override
public boolean onUnbind(Intent intent) {
    Log.i("MyService", " onUnbind ()");
    return super.onUnbind(intent);
}
//自定义一个接口,用于监听数据的改变
static interface CallBack{
    void dataChanged(String data);
}
//定义 Callback 的对象,以及 getter/setter 的方法
CallBack callback;
public CallBack getCallback() {
    return callback;
}
public void setCallback(CallBack callback) {
    this.callback = callback;
}
}
```

### 3. 注册 Service

在完成了 Service 的编码以后,需要在清单文件中注册 Serivce。清单文件中注册 Service 的方式与前面所讲的一致,只需要修改类名即可。

### 4. 编写界面交互代码(MainActivity)

MainActivity 主要负责实现 5 个按钮的功能,单击"启动服务"或"绑定服务"按钮可以将 EditText 中输入的数据传送到 MyService;在绑定服务之后,单击"同步服务数据"按钮,可以将 MyService 中修改后的数据显示到 TextView 中。具体代码如下所示:

```java
public class MainActivity extends Activity implements View.OnClickListener, ServiceConnection
{
    EditText etInput;
    TextView tvOut;
    Button btnStartService,btnStopService,btnBindService,btnUnbindService,btnSync;
    Intent serviceIntent;
    MyService.MyBinder binder = null;
    @Override
    protected void onCreate(Bundle savedInstanceState) {
        super.onCreate(savedInstanceState);
        setContentView(R.layout.activity_main);
        serviceIntent = new Intent(MainActivity.this,MyService.class);
        initCompent();                          //初始化组件
        btnClickListener();                     //给5个按钮添加事件监听
    }
    private void initCompent() {
        etInput = findViewById(R.id.etInput);
        tvOut = findViewById(R.id.tvOut);
        btnStartService = findViewById(R.id.btnStartService);
        btnStopService = findViewById(R.id.btnStopService);
        btnBindService = findViewById(R.id.btnBindService);
        btnUnbindService = findViewById(R.id.btnUnBindService);
        btnSync = findViewById(R.id.btnSync);
    }
    private void btnClickListener() {
        btnStartService.setOnClickListener(this);
        btnStopService.setOnClickListener(this);
        btnBindService.setOnClickListener(this);
        btnUnbindService.setOnClickListener(this);
        btnSync.setOnClickListener(this);
    }
    @Override
    public void onClick(View view) {
        switch (view.getId()){
            case R.id.btnStartService:
                serviceIntent.putExtra("data",etInput.getText().toString());
                startService(serviceIntent);        //启动服务传递参数
                break;
            case R.id.btnStopService:
                stopService(serviceIntent);
                break;
            case R.id.btnBindService:
                bindService(serviceIntent,this, Context.BIND_AUTO_CREATE);
                break;
            case R.id.btnUnBindService:
                unbindService(this);
                break;
            case R.id.btnSync:
                if(binder!= null){                  //绑定服务后,同步服务数据
```

```
                binder.setData(etInput.getText().toString());
            }
            break;
    }
}
@Override
public void onServiceConnected(ComponentName componentName, IBinder iBinder) {
    binder = (MyService.MyBinder) iBinder;
    binder.getService().setCallback(new MyService.CallBack() {
        @Override
        public void dataChanged(String data) {
            Message msg = new Message();
            Bundle b = new Bundle();
            b.putString("data",data);
            msg.setData(b);
            handler.sendMessage(msg);
        }
    });
}
@Override
public void onServiceDisconnected(ComponentName componentName) {
}
Handler handler = new Handler(){
    @Override
    public void handleMessage(Message msg) {
        super.handleMessage(msg);
        tvOut.setText(msg.getData().getString("data"));
    }
};
}
```

## 5. 运行程序

运行上述代码，首先单击用户界面中的"启动服务"按钮，EditText 中默认设置信息为 input message，通过 serviceIntent 传递到 MyService 中，打印的 Log 日志信息如图 8-11 所示。

```
2020-02-09 12:00:46.814 13272-13272/com.jxust.cn.chapter8_service_test I/MyService:  onCreate ()
2020-02-09 12:00:46.815 13272-13358/com.jxust.cn.chapter8_service_test I/MyService: 1:来自服务器的信息...
2020-02-09 12:00:46.815 13272-13272/com.jxust.cn.chapter8_service_test I/MyService:  onStartCommand ()
2020-02-09 12:00:47.816 13272-13358/com.jxust.cn.chapter8_service_test I/MyService: 2:input message
2020-02-09 12:00:48.820 13272-13358/com.jxust.cn.chapter8_service_test I/MyService: 3:input message
2020-02-09 12:00:49.822 13272-13358/com.jxust.cn.chapter8_service_test I/MyService: 4:input message
2020-02-09 12:00:50.825 13272-13358/com.jxust.cn.chapter8_service_test I/MyService: 5:input message
2020-02-09 12:00:51.827 13272-13358/com.jxust.cn.chapter8_service_test I/MyService: 6:input message
```

图 8-11 打印的 Log 日志信息

单击"停止服务"按钮以后，打印的 Log 信息如图 8-12 所示。

接下来测试绑定服务。单击"绑定服务"按钮，显示的结果如图 8-13 所示。

```
2020-02-09 12:00:51.827 13272-13358/com.jxust.cn.chapter8_service_test I/MyService: 6:input message
2020-02-09 12:00:52.830 13272-13358/com.jxust.cn.chapter8_service_test I/MyService: 7:input message
2020-02-09 12:00:53.832 13272-13358/com.jxust.cn.chapter8_service_test I/MyService: 8:input message
2020-02-09 12:00:54.834 13272-13358/com.jxust.cn.chapter8_service_test I/MyService: 9:input message
2020-02-09 12:00:55.836 13272-13358/com.jxust.cn.chapter8_service_test I/MyService: 10:input message
2020-02-09 12:00:56.724 13272-13272/com.jxust.cn.chapter8_service_test I/MyService:   onDestroy ()
```

图 8-12 "停止服务"结果

```
2020-02-09 12:09:08.154 13272-13272/com.jxust.cn.chapter8_service_test I/MyService:   onCreate ()
2020-02-09 12:09:08.157 13272-13272/com.jxust.cn.chapter8_service_test I/MyService: onBind()
2020-02-09 12:09:08.158 13272-13463/com.jxust.cn.chapter8_service_test I/MyService: 1:来自服务器的信息...
2020-02-09 12:09:09.160 13272-13463/com.jxust.cn.chapter8_service_test I/MyService: 2:来自服务器的信息...
2020-02-09 12:09:10.163 13272-13463/com.jxust.cn.chapter8_service_test I/MyService: 3:来自服务器的信息...
2020-02-09 12:09:11.166 13272-13463/com.jxust.cn.chapter8_service_test I/MyService: 4:来自服务器的信息...
```

图 8-13 "绑定服务"结果

再单击"同步服务数据"按钮,手机界面的运行结果如图 8-14 所示,Log 日志显示如图 8-15 所示。

图 8-14 "同步服务数据"结果

```
2020-02-09 12:13:30.123 13272-13272/com.jxust.cn.chapter8_service_test I/MyService:   onCreate ()
2020-02-09 12:13:30.124 13272-13272/com.jxust.cn.chapter8_service_test I/MyService: onBind()
2020-02-09 12:13:30.124 13272-13509/com.jxust.cn.chapter8_service_test I/MyService: 1:来自服务器的信息...
2020-02-09 12:13:31.125 13272-13509/com.jxust.cn.chapter8_service_test I/MyService: 2:来自服务器的信息...
2020-02-09 12:13:32.128 13272-13509/com.jxust.cn.chapter8_service_test I/MyService: 3:来自服务器的信息...
2020-02-09 12:13:33.130 13272-13509/com.jxust.cn.chapter8_service_test I/MyService: 4:来自服务器的信息...
2020-02-09 12:13:34.133 13272-13509/com.jxust.cn.chapter8_service_test I/MyService: 5:input message
2020-02-09 12:13:35.137 13272-13509/com.jxust.cn.chapter8_service_test I/MyService: 6:input message
2020-02-09 12:13:36.139 13272-13509/com.jxust.cn.chapter8_service_test I/MyService: 7:input message
```

图 8-15 "同步服务数据"的 Log 信息

单击"解除绑定服务"按钮以后,Log 日志打印的信息如图 8-16 所示。

以上就是一个带参数传递的启动和绑定本地服务并与之通信的一个简单实例代码,开发者可以使用本地服务做更复杂的开发,原理都是使用 IBinder 对象与 ServiceConnection 对象进行通信。

```
2020-02-09 12:16:26.597 13272-13593/com.jxust.cn.chapter8_service_test I/MyService: 6:input message
2020-02-09 12:16:27.603 13272-13593/com.jxust.cn.chapter8_service_test I/MyService: 7:input message
2020-02-09 12:16:28.608 13272-13593/com.jxust.cn.chapter8_service_test I/MyService: 8:input message
2020-02-09 12:16:29.609 13272-13593/com.jxust.cn.chapter8_service_test I/MyService: 9:input message
2020-02-09 12:16:30.614 13272-13593/com.jxust.cn.chapter8_service_test I/MyService: 10:input message
2020-02-09 12:16:31.176 13272-13272/com.jxust.cn.chapter8_service_test I/MyService:   onUnbind ()
2020-02-09 12:16:31.176 13272-13272/com.jxust.cn.chapter8_service_test I/MyService:   onDestroy ()
```

图 8-16 "解除绑定服务"的 Log 信息

### 8.3.3 跨应用启动服务实例

视频讲解

8.3.1 节介绍了远程服务通信主要是通过 AIDL 实现不同应用程序之间的通信,接下来将通过一个具体的实例说明如何实现跨应用启动服务。与 7.3.3 节的跨应用数据共享案例类似,在同一个项目中创建两个 Module:app 和 app2,用于模拟两个应用程序。app 定义服务,app2 负责启动和停止 app 的服务并传递数据。具体的步骤如下。

**1. 创建 Chapter8_ServiceAIDL 项目**

创建完项目后,当前默认的 Module 为 app,首先修改 app 的布局文件 activity_main.xml,把布局文件修改为线性布局方式,可修改 Textview 的 text 为 This is app Service(也可以不改,此处只用于显示)。

activity.xml 文件的具体代码如下所示:

```xml
<?xml version = "1.0" encoding = "utf-8"?>
< LinearLayout xmlns:android = "http://schemas.android.com/apk/res/android"
    xmlns:app = "http://schemas.android.com/apk/res-auto"
    xmlns:tools = "http://schemas.android.com/tools"
    android:layout_width = "match_parent"
    android:layout_height = "match_parent"
    android:orientation = "vertical"
    tools:context = ".MainActivity">
    < TextView
        android:layout_width = "wrap_content"
        android:layout_height = "wrap_content"
        android:layout_marginTop = "50dp"
        android:text = "This is APP Service"
        android:layout_gravity = "center" />
</LinearLayout >
```

**2. 创建 MyAppService 和 AIDL**

创建一个 Service 的子类 MyAppService,默认只有 OnBind()方法,添加 onCreate()和 onDestroy()方法。跨应用启动和停止服务可以通过 Intent 的 setComponent()方法实现,跨应用绑定服务则需要通过 AIDL 实现。下面介绍 AIDL 的创建过程。

(1) 选择 app,右击,在弹出的快捷菜单中选择 New→AIDL→AIDL File,如图 8-17 所示。

(2) 单击 AIDL File,默认名为 IMyAidlInterface,可以修改文件名,创建成功后在 app 下出现

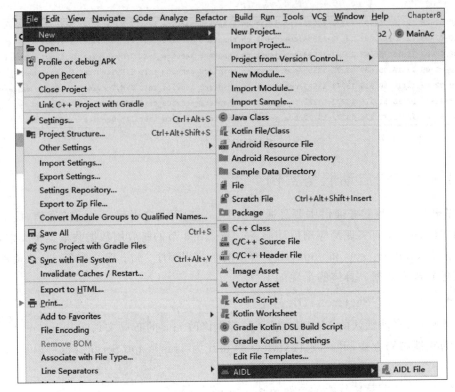

图 8-17 创建 AIDL 文件

aidl 文件夹及 IMyAidlInterface.aidl，如图 8-18 所示。

图 8-18 AIDL 文件

（3）AIDL 属于接口，包含一个抽象方法 basicTypes()，自定义添加一个 setData() 方法用于 app2 与 MyAppService 之间的通信。IMyAidlInterface 的代码如下所示：

```
interface IMyAidlInterface {
    void basicTypes(int anInt, long aLong, boolean aBoolean, float aFloat,double aDouble, String aString);
    void setData(String data);
}
```

（4）创建 AIDL 之后需要编译 IMyAidlInterface，然后修改 MyAppService 中的 onBind() 方法，返回 IMyAidlInterface.Stub() 的一个对象。实现 IMyAidlInterface 中 basicTypes() 和 setData() 两个方法。

（5）MyAppService 的代码如下所示：

```
public class MyAppService extends Service {
    private String data = "default value from MyAppService...";
    private boolean running = false;
    public MyAppService() {
    }
```

```java
    @Override
    public IBinder onBind(Intent intent) {
        return new IMyAidlInterface.Stub() {
            @Override
            public void basicTypes(int anInt, long aLong, boolean aBoolean, float aFloat, double
                            aDouble, String aString) throws RemoteException {
            }
            public void setData(String data) throws RemoteException {
                MyAppService.this.data = data;
            }
        };
    }
    @Override
    public void onCreate() {
        super.onCreate();
        Log.i("ServiceAIDL","Service onCreate");
        new Thread(){
            @Override
            public void run() {
                super.run();
                running = true;
                while(running){
                    try {
                        Log.i("ServiceAIDL","MyAppServive Data = " + data);
                        sleep(1000);
                    } catch (InterruptedException e) {
                        e.printStackTrace();
                    }
                }
            }
        }.start();
    }
    @Override
    public void onDestroy() {
        super.onDestroy();
        Log.i("ServiceAIDL","Service onDestroy");
        running = false;
    }
}
```

### 3. 修改 MainActivity

MainActivity 只需要显示界面，不需修改。具体代码如下所示：

```java
public class MainActivity extends AppCompatActivity {
    @Override
    protected void onCreate(Bundle savedInstanceState) {
        super.onCreate(savedInstanceState);
        setContentView(R.layout.activity_main);
    }
}
```

### 4. 创建 app2

在项目中创建 app2 启动 MyAppService。app2 的创建方法与 7.3.3 节中相同，不再赘述。修改 app2 中的布局和 Activity 分别为 activity_main2.xml 和 Main_Activity2。

activity_main2.xml 包含一个 EditText 和 5 个按钮。Android 5.0 以后 button 的属性 textAllCaps 的默认值为 true，即运行时会自动变成全大写，因而若要区分大小写需再改为 false。布局代码如下所示：

```xml
<?xml version = "1.0" encoding = "utf-8"?>
<LinearLayout xmlns:android = "http://schemas.android.com/apk/res/android"
    xmlns:app = "http://schemas.android.com/apk/res-auto"
    xmlns:tools = "http://schemas.android.com/tools"
    android:layout_width = "match_parent"
    android:layout_height = "match_parent"
    android:layout_marginTop = "16dp"
    android:orientation = "vertical"
    tools:context = ".MainActivity2">
    <EditText
        android:layout_width = "match_parent"
        android:layout_height = "wrap_content"
        android:text = "这是来自 app2 的信息"
        android:textStyle = "bold"
        android:textSize = "16dp"
        android:id = "@+id/etInput"/>
    <Button
        android:text = "启动 MyAppService 服务"
        android:layout_width = "match_parent"
        android:layout_height = "wrap_content"
        android:textAllCaps = "false"
        android:textSize = "16dp"
        android:id = "@+id/btnStartService"/>
    <Button
        android:text = "停止 MyAppService 服务"
        android:layout_width = "match_parent"
        android:layout_height = "wrap_content"
        android:textAllCaps = "false"
        android:textSize = "16dp"
        android:id = "@+id/btnStopService"/>
    <Button
        android:text = "绑定 MyAppService 服务"
        android:layout_width = "match_parent"
        android:layout_height = "wrap_content"
        android:textAllCaps = "false"
        android:textSize = "16dp"
        android:id = "@+id/btnBindService"/>
    <Button
        android:text = "解除绑定 MyAppService 服务"
        android:layout_width = "match_parent"
        android:layout_height = "wrap_content"
        android:textAllCaps = "false"
```

```
        android:textSize = "16dp"
        android:id = "@ + id/btnUnbindService"/>
    <Button
        android:text = "同步 MyAppService 数据"
        android:layout_width = "match_parent"
        android:layout_height = "wrap_content"
        android:textAllCaps = "false"
        android:textSize = "16dp"
        android:id = "@ + id/btnSync"/>
</LinearLayout>
```

### 5. 在 app2 中创建 AIDL

为了实现跨应用绑定服务后的数据通信,需要在 app2 中建立和 app 相同的 AIDL,即 app2 的 AIDL 与 app 中的 AIDL 包名和类名均相同,因而创建时选择另外一种方式。选中 app2,右击,在弹出的快捷菜单中选择 New→Folder→AIDL Folder,如图 8-19 所示。

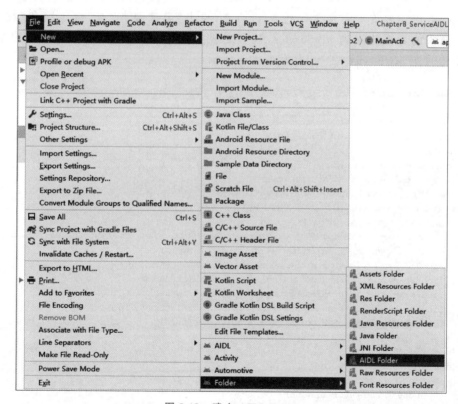

图 8-19 建立 AIDL Folder

在 app2 的 aidl 下,创建 package,复制 app 的包名并修改 package 的名称,之后将 app 的 IMyAidlInterface.aidl 文件复制到包下。结果如图 8-20 所示。

**注意**:app2 的 AIDL 建立之后,需要重新构建(Rebuild) app2,成功之后才可以在 MainActivity2 中调用 IMyAidlInterface。

图 8-20 建立与 app 相同的 AIDL

### 6. 修改 MainActivity2 代码

MainActivity2 主要负责实现 5 个按钮的功能,单击"启动服务"或"绑定服务"按钮可以将 EditText 中输入的数据传送到 MyService,在绑定服务之后,单击"同步服务数据"按钮,可以将 app2 的数据传递到 MyService 中。具体代码如下所示:

```java
public class MainActivity2 extends AppCompatActivity implements View.OnClickListener, ServiceConnection {
    private Intent serviceIntent;
    private EditText etInput;
    private IMyAidlInterface binder = null;
    @Override
    protected void onCreate(Bundle savedInstanceState) {
        super.onCreate(savedInstanceState);
        setContentView(R.layout.activity_main2);
        etInput = (EditText) findViewById(R.id.etInput);
        serviceIntent = new Intent();
        serviceIntent.setComponent(new ComponentName("com.jxust.cn.chapter8_serviceaidl",
                "com.jxust.cn.chapter8_serviceaidl.MyAppService"));
        findViewById(R.id.btnStartService).setOnClickListener(this);
        findViewById(R.id.btnStopService).setOnClickListener(this);
        findViewById(R.id.btnBindService).setOnClickListener(this);
        findViewById(R.id.btnUnbindService).setOnClickListener(this);
        findViewById(R.id.btnSync).setOnClickListener(this);
    }
    @Override
    public void onClick(View view) {
        switch (view.getId()){
            case R.id.btnStartService:
                startService(serviceIntent);
                break;
            case R.id.btnStopService:
                stopService(serviceIntent);
                break;
            case R.id.btnBindService:
                bindService(serviceIntent,this, Context.BIND_AUTO_CREATE);
```

```
                break;
            case R.id.btnUnbindService:
                unbindService(this);
                binder = null;
                break;
            case R.id.btnSync:
                if(binder!= null){
                    try {
                        binder.setData(etInput.getText().toString());
                    } catch (RemoteException e) {
                        e.printStackTrace();
                    }
                }
                break;
        }
    }
    @Override
    public void onServiceConnected(ComponentName componentName, IBinder iBinder) {
        binder = IMyAidlInterface.Stub.asInterface(iBinder);
    }
    @Override
    public void onServiceDisconnected(ComponentName componentName) {
    }
}
```

### 7. 运行程序

上述代码完成以后，运行程序。因为同一个项目中两个Module，运行时注意选择。与7.3.3节类似，首先运行app，然后运行app2，运行界面分别如图8-21和图8-22所示。

图8-21　app 运行界面

图8-22　app2 运行界面

在app2中单击用户界面中的"启动MyAppService服务"按钮，启动了MyAppService，打印的Log日志信息如图8-23所示。

单击"停止MyAppService服务"按钮，打印的Log信息如图8-24所示。

```
20042-20042/com.jxust.cn.chapter8_serviceaidl I/ServiceAIDL: Service onCreate
20042-20149/com.jxust.cn.chapter8_serviceaidl I/ServiceAIDL: MyAppServive Data =default value from MyAppService...
20042-20149/com.jxust.cn.chapter8_serviceaidl I/ServiceAIDL: MyAppServive Data =default value from MyAppService...
20042-20149/com.jxust.cn.chapter8_serviceaidl I/ServiceAIDL: MyAppServive Data =default value from MyAppService...
20042-20149/com.jxust.cn.chapter8_serviceaidl I/ServiceAIDL: MyAppServive Data =default value from MyAppService...
```

图 8-23 "启动 MyAppService 服务"结果

```
20042-20149/com.jxust.cn.chapter8_serviceaidl I/ServiceAIDL: MyAppServive Data =default value from MyAppService...
20042-20149/com.jxust.cn.chapter8_serviceaidl I/ServiceAIDL: MyAppServive Data =default value from MyAppService...
20042-20149/com.jxust.cn.chapter8_serviceaidl I/ServiceAIDL: MyAppServive Data =default value from MyAppService...
20042-20149/com.jxust.cn.chapter8_serviceaidl I/ServiceAIDL: MyAppServive Data =default value from MyAppService...
20042-20042/com.jxust.cn.chapter8_serviceaidl I/ServiceAIDL: Service onDestroy
```

图 8-24 "停止 MyAppService 服务"结果

接下来测试绑定服务。单击"绑定 MyAppService 服务"按钮，显示的结果如图 8-25 所示。

```
20042-20042/com.jxust.cn.chapter8_serviceaidl I/ServiceAIDL: Service onCreate
20042-20216/com.jxust.cn.chapter8_serviceaidl I/ServiceAIDL: MyAppServive Data =default value from MyAppService...
20042-20216/com.jxust.cn.chapter8_serviceaidl I/ServiceAIDL: MyAppServive Data =default value from MyAppService...
20042-20216/com.jxust.cn.chapter8_serviceaidl I/ServiceAIDL: MyAppServive Data =default value from MyAppService...
20042-20216/com.jxust.cn.chapter8_serviceaidl I/ServiceAIDL: MyAppServive Data =default value from MyAppService...
```

图 8-25 "绑定 MyAppService 服务"结果

再单击"同步 MyAppService 数据"按钮，Log 日志显示如图 8-26 所示。

```
20042-20216/com.jxust.cn.chapter8_serviceaidl I/ServiceAIDL: MyAppServive Data =default value from MyAppService...
20042-20216/com.jxust.cn.chapter8_serviceaidl I/ServiceAIDL: MyAppServive Data =这是来自app2的信息
20042-20216/com.jxust.cn.chapter8_serviceaidl I/ServiceAIDL: MyAppServive Data =这是来自app2的信息
20042-20216/com.jxust.cn.chapter8_serviceaidl I/ServiceAIDL: MyAppServive Data =这是来自app2的信息
20042-20216/com.jxust.cn.chapter8_serviceaidl I/ServiceAIDL: MyAppServive Data =这是来自app2的信息
```

图 8-26 "同步 MyAppService 数据"的 Log 信息

单击"解除绑定 MyAppService 服务"按钮以后，打印的 Log 信息如图 8-27 所示。

```
20042-20216/com.jxust.cn.chapter8_serviceaidl I/ServiceAIDL: MyAppServive Data =这是来自app2的信息
20042-20216/com.jxust.cn.chapter8_serviceaidl I/ServiceAIDL: MyAppServive Data =这是来自app2的信息
20042-20216/com.jxust.cn.chapter8_serviceaidl I/ServiceAIDL: MyAppServive Data =这是来自app2的信息
20042-20042/com.jxust.cn.chapter8_serviceaidl I/ServiceAIDL: Service onDestroy
```

图 8-27 "解除绑定 MyAppService 服务"的 Log 信息

## 8.4 系统服务类的使用

### 8.4.1 TelephonyManager

视频讲解

TelephonyManager(电话管理器)是一个管理手机通话状态、电话网络信息的服务类，该类提供了大量的 getXxx() 方法来获取电话网络相关的信息。

在程序中获取 TelephonyManager 很容易,只要调用如下所示的代码即可:

```
//获取系统的 TelephonyManager 对象
TelephonyManager telManager =
(TelephonyManager)getSystemService(Context.TELEPHONY_SERVICE);
```

接下来就可以使用 TelephonyManager 获取相关信息和进行相关操作了。

下面将通过一个实例——获取网络和 SIM 卡信息来讲解电话管理器的使用,具体的步骤如下。

### 1. 创建 Chapter8_TelephonyManager 项目

创建完项目以后,在布局界面中添加一个 ListView 组件,用来显示获取到的信息,布局界面在这里就不再赘述。

### 2. 编写 MainActivity 代码

MainActivity 类主要负责获取系统的 TelephonyManager 对象、初始化 ListView 组件、数据加载到 ListView。MainActivity 的具体代码如下所示:

```java
public class MainActivity extends AppCompatActivity {
    private ListView show_list;
    private ArrayList<String> status_values = new ArrayList<String>();
    @Override
    protected void onCreate(Bundle savedInstanceState) {
        super.onCreate(savedInstanceState);
        setContentView(R.layout.activity_main);
        show_list = (ListView)findViewById(R.id.show_mes);
        //获取系统的 TelephonyManager 对象
        TelephonyManager telManager =
            (TelephonyManager)getSystemService(Context.TELEPHONY_SERVICE);
        //获取设备编号
        status_values.add(telManager.getDeviceId());
        //获取系统平台的版本
        status_values.add(telManager.getDeviceSoftwareVersion()
                != null? telManager.getDeviceSoftwareVersion():"未知");
        //获取网络运营商代号
        status_values.add(telManager.getNetworkOperator());
        //获取 SIM 卡的国别
        status_values.add(telManager.getSimCountryIso());
        //获取 SIM 卡的序列号
        status_values.add(telManager.getSimSerialNumber());
        show_list.setAdapter(new ArrayAdapter<String>(this,
                android.R.layout.simple_list_item_1, status_values));
    }
}
```

### 3. 添加读取权限

由于该应用需要获取手机状态信息,因此需要在清单文件中添加权限。具体代码如下所示:

```
<uses-permission android:name="android.permission.READ_PHONE_STATE"/>
```

### 4. 运行程序

完成了代码编写和权限添加后,就可以运行程序。首次加载程序后,需要在"设置"中打开权限,具体操作与 6.2.1 节中图 6-1 的 SD 卡权限相类似,为模拟器打开当前项目读取 Phone 的权限,如图 8-28 所示。再次运行程序,结果如图 8-29 所示。

图 8-28 打开读取 Phone 权限

图 8-29 读取手机和 SIM 卡信息的结果

TelephonyManager 除了提供一系列的 getXxx()方法来获取网络和 SIM 卡信息,还提供了一个 listen(PhoneStateListener listener,int events)方法来监听通话状态。学习者可以自己创建一个项目,了解执行过程。

## 8.4.2 SmsManager

SmsManager(短信管理器)是 Android 提供的一个非常常见的服务,它提供了一系列的 sendXxxMessage()方法用于发送短信。短信通常都是文本的形式,通过调用 sendTextMessage()方法即可实现。

接下来将通过一个发送短信的案例来讲解 SmsManager 的使用。本例非常简单,提供两个输入框,分别用来输入号码和内容,然后一个按钮用来发送。具体步骤如下。

### 1. 创建 Chapter8_SmsManager 项目

修改布局文件 activity_main.xml 的代码,采用线性垂直布局的方式,放置两个 EditText 组件和一个按钮组件。具体代码如下所示:

```
<?xml version="1.0" encoding="utf-8"?>
<LinearLayout xmlns:android="http://schemas.android.com/apk/res/android"
    xmlns:tools="http://schemas.android.com/tools"
    android:layout_width="match_parent"
    android:layout_height="match_parent"
    tools:context="com.jxust.cn.chapter8_smsmanager.MainActivity"
    android:orientation="vertical">
    <EditText
```

```xml
        android:id = "@+id/number"
        android:layout_width = "match_parent"
        android:layout_height = "wrap_content"
        android:hint = "输入号码"/>
    <EditText
        android:id = "@+id/content"
        android:layout_width = "match_parent"
        android:layout_height = "wrap_content"
        android:hint = "输入内容"/>
    <Button
        android:id = "@+id/send"
        android:layout_width = "match_parent"
        android:layout_height = "wrap_content"
        android:text = "发送"/>
</LinearLayout>
```

### 2. 编写用户交互界面代码（MainActivity）

MainActivity 负责初始化 3 个组件，然后调用短信发送类 SmsManager 的 sendTextMessage() 来发送短信。具体代码如下所示：

```java
public class MainActivity extends AppCompatActivity {
    private EditText number,content;
    private Button send;
    SmsManager smsManager;
    protected void onCreate(Bundle savedInstanceState) {
        super.onCreate(savedInstanceState);
        setContentView(R.layout.activity_main);
        //获取 SmsManager
        smsManager = SmsManager.getDefault();
        //初始化组件
        number = (EditText)findViewById(R.id.number);
        content = (EditText)findViewById(R.id.content);
        send = (Button)findViewById(R.id.send);
        send.setOnClickListener(new View.OnClickListener() {
            public void onClick(View view) {
                //创建一个 PendingIntent 对象
PendingIntent pintent = PendingIntent.getActivity(MainActivity.this, 0, new Intent(), 0);
                //发送短信
        smsManager.sendTextMessage(number.getText().toString() ,
                        null, content.getText().toString(), pintent,null);
        Toast.makeText(MainActivity.this, "发送成功", Toast.LENGTH_SHORT).show();
            }
        });
    }
}
```

上面代码中，使用到了一个 PendingIntent 对象，PendingIntent 是对 Intent 的一种包装，一般通过调用 PendingIntent 对象的 getActivity()、getService()等静态方法来获取 PendingIntent 对象。

### 3. 添加权限

因为程序需要调用 SmsManager 来发送短信,因此还需要对该程序添加发送短信的权限。具体代码如下所示。

```
<uses-permission android:name="android.permission.SEND_SMS"/>
```

图 8-30　打开读取 SMS 权限

### 4. 运行程序

完成了代码编写和权限添加后,就可以运行程序。首次加载程序后,需要在"设置"中打开权限,具体操作与 6.2.1 节中图 6-1 的 SD 卡权限相类似,为模拟器打开当前项目读取 SMS 的权限,如图 8-30 所示。再次运行程序,输入模拟器中联系人的号码以及短信内容,单击"发送"按钮,结果如图 8-31 所示。发送成功后,在短信界面查询,可以看到该联系人收到了短信,如图 8-32 所示。

图 8-31　发送短信结果

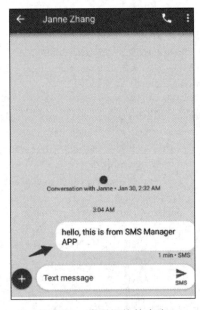

图 8-32　收到短信的内容

## 8.5　广播消息

### 8.5.1　广播简介

在 Android 系统中,有一些操作完成以后,会发送广播,比如发出短信或打出一个电话。如果某个程序接收到这个广播,就会做出相应的处理。因为它只负责发送消息,而不管接收方如何处理,所以叫它广播。广播可以被多个应用程序接收,也可以不被任何应用程序接收。

BroadcastReceiver 是负责对发送出来的广播进行过滤接收并响应的一类组件。BroadcastReceiver 和事件处理机制类似,不同的是广播事件处理机制是系统级别的,而事件处理机制是应用程序组件级别的。

接下来将详细介绍 Broadcast 和 BroadcastReceiver 的使用。

首先在需要发送短信的地方,把要发送和用于过滤的信息装入一个 Intent 对象,然后通过调用 Context.sendBroadcast()、sendOrderBroadcast()或者 sendStickyBroadcast()方法,将 Intent 对象以广播的方式发送出去。

当 Intent 发送以后,所有已经注册的 BroadcastReceiver 会检查注册时的 IntentFilter 是否与发送的 Intent 相匹配,若匹配则会调用 BroadcastReceiver 的 onReceive()方法。需要注意的是,在定义 BroadcastReceiver 时,需要实现它的 onReceive()方法。

广播的两种注册方式如下。

- 静态地在清单文件中使用<receiver>标签进行注册,并在标签内使用<intent-filter>标签设置过滤器。
- 动态地在代码中先定义并设置好一个 IntentFilter 对象,然后在需要注册的地方调用 Context.registerReceiver()方法,如果取消注册时就调用 Context.unregisterReceiver()方法。动态方式注册的广播,如果它的 Context 对象被销毁,BroadcastReceiver 也就会自动取消注册。

接下来将会通过具体的例子来讲解这两种注册方式的使用过程。

## 8.5.2 静态注册广播案例

本案例采用静态注册的方式来发送广播,布局相对简单,界面上放置一个按钮,单击按钮以后就会发送一个广播,当广播接收器收到该广播时就会在界面弹出一个提示消息。具体操作步骤如下。

视频讲解

### 1. 创建 Chapter8_broadcast_static 项目

修改布局界面代码,添加一个按钮组件。activity_main.xml 文件的具体代码如下所示:

```xml
<?xml version = "1.0" encoding = "utf-8"?>
<LinearLayout xmlns:android = "http://schemas.android.com/apk/res/android"
    xmlns:tools = "http://schemas.android.com/tools"
    android:layout_width = "match_parent"
    android:layout_height = "match_parent"
    tools:context = "com.jxust.cn.chapter8_broadcast_static.MainActivity"
    android:orientation = "vertical">
<Button
    android:id = "@ + id/button"
    android:layout_width = "match_parent"
    android:layout_height = "wrap_content"
    android:text = "发送广播" />
</LinearLayout>
```

## 2. 编写 MainActivity 代码

MainActivity 主要负责初始化按钮组件，然后添加单击事件，发送广播。具体代码如下所示：

```java
public class MainActivity extends AppCompatActivity {
    private Button send_btn;
    //此值与对应的 Receiver 里的过滤器的值相同
    private final String action = "MyBroadcast";
    @Override
    protected void onCreate(Bundle savedInstanceState) {
        super.onCreate(savedInstanceState);
        setContentView(R.layout.activity_main);
        send_btn = (Button)findViewById(R.id.button);
        //设置监听
        send_btn.setOnClickListener(new View.OnClickListener() {
            @Override
            public void onClick(View view) {
                Intent intent = new Intent();
                intent.setPackage("com.jxust.cn.chapter8_broadcast_static");
                intent.setAction(action);
                //发送广播
                MainActivity.this.sendBroadcast(intent);
            }
        });
    }
}
```

上述代码为按钮添加了单击事件，当按钮按下时，就会发送广播。其中 package 为当前项目的包名，action 的值与清单文件中定义的值相同。

## 3. 自定义广播接收器

自定义广播接收器继承自 BroadcastReceiver，然后实现它的 onReceive() 方法。具体代码如下所示：

```java
public class MyReceiver extends BroadcastReceiver {
    @Override
    public void onReceive(Context context, Intent intent) {
        //收到广播时,显示一个通知
        Toast.makeText(context,"广播接收成功",Toast.LENGTH_SHORT).show();
    }
}
```

从上述代码中可以看出，当接收广播时，就会在主界面上显示一个内容为"广播接收成功"的通知。

## 4. 注册广播接收器

由于本案例使用的是静态的注册方式，所以需要在 AndroidManifest.xml 文件中注册。具体代码如下所示：

```xml
<receiver android:name=".MyReceiver">
    <intent-filter>
        <action android:name="MyBroadcast" />
    </intent-filter>
</receiver>
```

从上述代码中可以看出，</intent-filter>里定义的 action 的 name 属性值要与发送广播时的字符串相同。

**5. 运行程序**

上述操作完成以后，运行程序，单击主界面上的"发送广播"按钮，就会弹出一个通知，具体结果如图 8-33 所示。

### 8.5.3 动态注册广播案例

动态注册广播要删除清单文件中关于 Receiver 的定义在代码中实现。具体操作过程如下。

**1. 创建 Chapter8_Broadcast_dynamic 项目**

修改布局界面的代码，采用线性布局的方式，放置 3 个按钮组件，分别是"注册广播接收器""发送广播"和"取消广播接收器"。activity_main.xml 的具体代码如下所示：

图 8-33 "静态注册广播"结果

```xml
<?xml version="1.0" encoding="utf-8"?>
<LinearLayout xmlns:android="http://schemas.android.com/apk/res/android"
    xmlns:app="http://schemas.android.com/apk/res-auto"
    xmlns:tools="http://schemas.android.com/apk/res/tools"
    android:layout_width="match_parent"
    android:layout_height="match_parent"
    android:orientation="vertical"
    tools:context=".MainActivity">
    <Button
        android:layout_width="match_parent"
        android:layout_height="wrap_content"
        android:layout_marginTop="6dp"
        android:text="注册广播接收器"
        android:id="@+id/btnRegister"
        />
    <Button
        android:layout_width="match_parent"
        android:layout_height="wrap_content"
        android:layout_marginTop="6dp"
        android:text="发送广播"
        android:id="@+id/btnSend"
        />
    <Button
        android:layout_width="match_parent"
```

```
            android:layout_height = "wrap_content"
            android:layout_marginTop = "6dp"
            android:text = "取消注册广播接收器"
            android:id = "@ + id/btnUnRegister"
            />
</LinearLayout>
```

### 2. 编写 MainActivity 代码

MainActivity 负责初始化按钮组件、添加监听事件、设置广播接收者的 action 的 package 和 name 属性值，动态注册广播接收器，具体代码如下所示：

```java
public class MainActivity extends AppCompatActivity implements OnClickListener {
    private Button btnRegister,btnSend,btnUnRegister;
    MyReceiver receiver = null;
    @Override
    protected void onCreate(Bundle savedInstanceState) {
        super.onCreate(savedInstanceState);
        setContentView(R.layout.activity_main);
        btnRegister = findViewById(R.id.btnRegister);
        btnSend = findViewById(R.id.btnSend);
        btnUnRegister = findViewById(R.id.btnUnRegister);
        btnRegister.setOnClickListener(this);
        btnSend.setOnClickListener(this);
        btnUnRegister.setOnClickListener(this);
    }
    @Override
    public void onClick(View view) {
        switch (view.getId()){
            case R.id.btnRegister:
                if(receiver == null){
                    receiver = new MyReceiver();
                    registerReceiver(receiver,new IntentFilter(MyReceiver.ACTION));
                    Log.i("BroadcastTest","注册成功");
                }
                break;
            case R.id.btnSend:
                Intent intent = new Intent();
                intent.putExtra("data","this is a test");
                intent.setPackage("com.jxust.cn.chapter_broadcast_dynamic");
                intent.setAction(MyReceiver.ACTION);
                sendBroadcast(intent);
                break;
            case R.id.btnUnRegister:
                if(receiver!= null){
                    unregisterReceiver(receiver);
                    receiver = null;
                    Log.i("BroadcastTest","注销成功");
```

```
                }
                break;
        }
    }
}
```

该代码中,新建了一个 Receiver 对象和 IntentFilter 对象,并将其作为参数调用 registerReceiver() 注册方法。当单击按钮时,广播接收器才会被注册,此时才能检测到短信的接收消息。

**3. 自定义广播接收器**

自定义一个广播接收器继承自 BroadcastReceiver,当接收到消息时,显示一个通知。具体代码如下所示:

```
public class MyReceiver extends BroadcastReceiver {
    static final String ACTION = "com.jxust.cn.chapter_broadcast_dynamic.cast";
    @Override
    public void onReceive(Context context, Intent intent) {
        String s = "接收到广播,接收的信息是:" + intent.getStringExtra("data");
        Toast.makeText(context,s,Toast.LENGTH_SHORT).show();
        Log.i("BroadcastTest",s);
        abortBroadcast();
    }
}
```

**4. 运行程序**

运行程序,先单击"注册广播接收器"按钮注册广播接收器,然后单击"发送广播"按钮,则可以发送广播,结果如图 8-34 所示。若单击"取消注册广播接收器"按钮,再次单击"发送广播"按钮则不能再发送。具体的 Log 日志如图 8-35 所示。

图 8-34 "动态注册广播"结果

图 8-35 "动态注册广播"Log 日志

## 8.6 【项目案例 6】 "橙一色"App 消息推送功能实现

利用本章学习的知识，为"橙一色"App 添加消息推送通知的功能。即利用 Service 和 Android 消息通知 NotificationManager 类来定时通知当前时间。定义一个 Service 的子类在后台开启一个定时任务，每隔一分钟在上方通知栏显示一次通知，定时的功能使用 Android 定时器 Timer 类实现。

在 Android 客户端有时候有些任务不是当时就执行，而是指定一个特定时间的几秒之后执行任务，这个时候就需要用到定时器。开启一个定时器的代码如下所示：

```java
Timer timer = new Timer();
TimerTask task = new TimerTask() {
    @Override
    public void run() {
        try {
            holder.post(url,"",200);
        } catch (IOException e) {
            e.printStackTrace();
        }
    }
};
```

通过上述代码即可定义一个定时器，可以看到 run() 方法是向服务器发送 POST 请求来返回数据刷新界面。接下来就需要开启定时器任务。具体代码如下所示：

```java
timer.schedule(task,0,10000);
```

可以看出，上述代码中需要传递 3 个参数：第一个参数是要执行的定时任务；第二个参数是延迟时间，指的是多久以后才执行这个功能；第三个参数是间隔时间，指多久执行一次。

其实定时器如果不销毁的话是一直执行的，如果定时器一直执行，那么程序运行不了多长时间就会崩溃，所以定时器用完要及时关闭。定时器关闭一般写在 Fragment 或者 Activity 的 onStop() 方法中。具体代码如下：

```java
timer.cancel();
```

本节就利用上面的定时器，使用 Service 在后台开启一个服务，目前先定时推送当前时间，第 12 章优化中将它改造成定时向服务器发送请求，然后获取最新的推送消息。首先，定义 TimeService 继承自 Service。具体代码如下：

```java
public class TimeService extends Service {
    private static Timer timer = null;
    private NotificationManager manager;
    private NotificationCompat.Builder builder;
    public IBinder onBind(Intent intent) {
        return null;
    }
    //清除通知
    private void cleanAllNotification() {
        manager = (NotificationManager) getSystemService(Context.NOTIFICATION_SERVICE);
        manager.cancelAll();
        if (timer != null) {
            timer.cancel();
            timer = null;
        }
    }
    //添加通知
    public int onStartCommand(Intent intent,int flags,int startId) {
        long period = 60 * 1000; //1分钟一个周期
        if (timer == null) {
            timer = new Timer();
        }
        timer.schedule(new TimerTask() {
            @Override
            public void run() {
                /**
                 * NotificationChannel 是 Android 8.0 新增的特性,如果 App 的
                   targetSDKVersion >= 26,
                 * 没有设置 channel 通知渠道的话,就会导致通知无法展示。
                 * 报错内容:Failed to post notification on channel "null"
                 */
                NotificationChannel channel = new NotificationChannel("channel", "通知",
                                NotificationManager.IMPORTANCE_DEFAULT);
                manager = (NotificationManager)
                        getSystemService(Context.NOTIFICATION_SERVICE);
                manager.createNotificationChannel(channel);
                builder = new NotificationCompat.Builder(TimeService.this);
                builder.setContentTitle("提醒系统当前时间");
                builder.setSmallIcon(R.drawable.chengzi);
                builder.setChannelId("channel");
                Date date = new Date();
                SimpleDateFormat formatter = new SimpleDateFormat("yyyy-MM-dd HH:mm:ss");
                builder.setContentText("系统当前时间为:" + formatter.format(date));
                builder.setAutoCancel(true);
                builder.setDefaults(Notification.DEFAULT_ALL);
                Notification notification = builder.build();
                manager.notify(1,notification);
            }
        }, 0, period);
        return super.onStartCommand(intent, flags, startId);
    }
}
```

Service 也是 Android 中的四大组件之一,所以需要在 AndroidManifest.xml 中进行注册,本节设置 Service 中的属性 android:process=system,将其运行在单独的进程中,当系统退出时,依然可以运行。具体代码如下所示:

```xml
<?xml version = "1.0" encoding = "utf-8"?>
< manifest xmlns:android = "http://schemas.android.com/apk/res/android"
    package = "com.OrangeSale.cn">
    < application
        < activity android:name = ".MainActivity">
            < intent - filter >
                < action android:name = "android.intent.action.MAIN" />
                < category android:name = "android.intent.category.LAUNCHER" />
            </intent - filter >
        < service android:enabled = "true"
        android:name = ".service.TimeService" android:process = "system" />
    </application >
</manifest >
```

上面加粗代码即是注册自定义的 TimeService。

通过以上代码,即可执行每隔一分钟,定时通知一下当前时间,实现的具体效果分别如图 8-36 和图 8-37 所示。

图 8-36　系统通知 1

图 8-37　系统通知 2

从图 8-36 和图 8-37 即可看出,定时任务每隔一分钟执行通知下系统当前的时间,后续会实现一个前端界面定时推送消息,利用此 Service 定时去请求服务器获取推送的消息,达到消息推送的功能。

## 8.7　本章小结

本章主要讲解 Android 中的 Service 和 Broadcast。首先讲解服务的基本概念、创建与配置、启动与停止,接着讲解服务的生命周期所经历的过程、生命周期的方法,然后又讲解服务的通信,包括本地服务通信与远程服务通信,最后讲解系统服务的使用。广播方面做了简单的介绍,通过具体的例子讲解广播的注册及使用,最后实现了"橙一色"App 的消息推送功能。关于广播和服务的知识,需要开发者熟练掌握,在日常的开发中会经常用到。

## 8.8　课后习题

1. 简介 Service 的两种启动方式的特点。
2. 简介 Service 的生命周期。
3. 简介 Service 的两种通信方式的特点。
4. Broadcast 有哪几种不同的注册方式？有什么区别。
5. 编写程序,要求程序关闭一段时间后,重新启动该程序。

# 第 9 章

# Android 网络通信编程

**学习目标**
- 掌握 HTTP。
- 掌握 HttpURLConnection、HttpClient 的使用。
- 掌握 Socket 通信的使用。
- 掌握 GET、POST 两种数据提交方式。

网络的发展，使移动端拥有无限的发展可能，而 Android 系统最大的特色和优势之一即是对网络的支持。目前几乎所有的 Android 应用程序都会涉及网络编程。

Android 系统提供了 Socket 通信、HTTP 通信、URL 通信和 WebView。其中最常用的是 HTTP 通信，本章将会详细地讲解 HTTP 通信和 Socket 通信。

## 9.1 网络编程基础

### 9.1.1 HTTP 简介

HTTP 是 Hyper Text Transfer Protocol（超文本传输协议）的缩写，是用于从万维网（World Wide Web，WWW）服务器传输超文本到本地浏览器的传送协议。HTTP 是基于 TCP/IP 通信协议来传递数据（HTML 文件，图片文件，查询结果等）的。

HTTP 是一个属于应用层的面向对象的协议，由于其简捷、快速的方式，适用于分布式超媒体信息系统。它于 1990 年提出，经过几年的使用与发展，得到不断完善和扩展。目前在 WWW 中使用的是 HTTP 1.1 的第 6 版，HTTP 2.1 的规范化工作正在进行之中。

HTTP 工作于客户端-服务端架构上。浏览器作为 HTTP 客户端通过 URL 向 HTTP 服务端即 Web 服务器发送所有请求。Web 服务器根据接收到的请求，向客户端发送响应信息。

HTTP 请求到响应的过程如图 9-1 所示。

HTTP 的主要特点如下。

(1)简单快速:客户端向服务器请求服务时,只需传送请求方法和路径。请求方法常用的有 GET、HEAD、POST。

图 9-1  HTTP 请求-响应图

(2)灵活:HTTP 允许传输任意类型的数据对象。

(3)无连接:无连接的含义是限制每次连接只处理一个请求。服务器处理完客户端的请求,并收到客户端的应答后,即断开连接。

(4)无状态:HTTP 是无状态协议。无状态是指协议对于事务处理没有记忆能力。缺少状态意味着如果后续处理需要前面的信息,则它必须重传,这样可能导致每次连接传送的数据量增大。

(5)支持 B/S 及 C/S 模式。

## 9.1.2 标准 Java 接口

标准 Java 接口是指 java.net.*,它提供与互联网有关的类,包括流和数据包套接字、Internet 协议和一些 HTTP 处理。例如创建 URL 和 URLConnection 对象、设置连接参数、连接服务器等。java.net.* 提供的类以及接口如表 9-1 所示。

表 9-1  java.net.* 提供的类和接口说明

| 类/接口 | 说　　明 |
| --- | --- |
| ServerSocket | 实现服务器套接字 |
| Socket | 实现客户端套接字 |
| DatagramSocket | 表示用来发送和接收数据报的套接字 |
| InterAddress | 表示互联网协议(IP)地址 |
| HttpURLConnection | 用于管理 HTTP 链接的资源链接管理器 |
| URL | 代表一个统一资源定位符,它是指向互联网"资源"的指针 |

为了更好地讲解 java.net 包的 HTTP 的方法的使用,接下来将通过一段代码来说明。具体代码如下所示:

```java
try{
    //定义地址
    URL url = new URL("http://localhost:8080/Test/index.jsp");
    //打开链接地址
    HttpURLConnection http = (HttpURLConnection)url.openConnection();
    //得到连接状态
    int state = http.getResponseCode();
    if (state == HttpURLConnection.HTTP_OK) {
        //取得数据
        InputStream in = http.getInputStream();
        //处理数据
        ...
    }
}catch(Exception e){
```

```
        }
    }
```

以上就是 java.net 包的 HTTP 的方法应用。从中可以看出在建立连接以后，可以通过调用 HttpURLConnection 连接对象的 getInputStream() 函数，将内存缓冲区中封装好的完整的 HTTP 请求电文发送到服务端。

### 9.1.3 Android 网络接口

Android 的网络接口即是 android.net.*，它实际上是通过对 Apache HttpClient 的封装来实现一个 HTTP 编程接口，比 java.net.* API 功能更强大。android.net.* 除了具备核心的 java.net.* 外，还包含额外的网络访问 Socket。该包包括 URI 类，在 Android 应用程序开发中使用较多。同时 Android 网络接口还提供了 HTTP 请求队列管理、HTTP 连接池管理、网络状态监视等接口。

实现 Socket 的连接功能的具体代码如下所示：

```
try{
    //IP 地址定义
    InetAddress address = InetAddress.getByName("192.168.56.1");
    //端口定义
    Socket client = new Socket(address,"61111",true);
    //取得数据
    InputStream in = client.getInputStream();
    OutputStream out = client.getOutputStream();
    //处理数据
    ...
    out.close();
    in.close();
    client.close();
}catch(UnknownHostException e){

}
catch(Exception e){

}
```

## 9.2 HTTP 通信

### 9.2.1 HttpURLConnection 简介

在 Android 开发中，应用程序经常需要与服务器进行数据交互，包括访问本地服务器以及远程服务器，这些都可以称为访问网络，此时就可以使用 HttpURLConnection 对象。它是一个标准的

Java 类,在 9.1.2 节中已经讲解了它的用法。9.1.2 节的代码主要演示了手机端与服务器建立连接并获取服务器返回数据的过程。

HttpURLConnection 继承自 URLConnection 类,两者都是抽象类,其对象主要通过 URL 的 openConnection()方法获得。

openConnection()方法只创建 URLConnection 或者 HttpURLConnection 实例,但并不进行真正的连接操作,并且每次 openConnection()都将创建一个新的实例。因此在连接之前可以对它的一些属性进行设置。

设置超时时间,以及设置请求方式的具体代码如下所示:

```
//设置请求方式
http.setRequestMethod("GET");
//设置超时时间
http.setConnectionTimeout(4000);
```

需要注意的是在连接时需要设置超时时间,如果不设置超时时间,在网络异常的情况下,会导致取不到数据而一直等待,以至于程序不能继续往下执行。

在开发 Android 应用程序的过程中,如果应用程序需要访问网络权限,则需要在清单文件中添加如下所示的代码:

```
< uses - permission android:name = "android.permission.INTERNET" />
```

为保证用户数据和设备的安全,Google 公司针对 Android 6.0 以上版本的应用程序,要求默认使用加密连接,因此运行 Android 系统的安卓设备无论是接收或者发送流量,都不能明码传输,需要使用下一代传输层安全协议(Transport Layer Security)。

如果在高版本 Android 系统中使用 HttpURLConnection 直接进行 HTTP 请求,通常会报出以下异常:

```
W/System.err: java.io.IOException: Cleartext HTTP traffic to **** not permitted
```

若使用 OKHttp 或 HttpClient 请求,则报出以下异常:

```
java.net.UnknownServiceException: CLEARTEXT communication ** not permitted by network security policy
```

针对上述问题,有以下 4 种解决方法:

(1) 将 HTTP 请求改为 HTTPS 请求。
(2) targetSdkVersion 降到 27 以下。
(3) 更改网络安全配置。
(4) 直接在 AndroidManifest.xml 中更改配置。

前面两个方法容易理解和实现,具体说明第三种方法——更改网络安全配置,方法如下。

(1) 在 res 文件夹下创建一个 xml 文件夹,然后创建一个 network_security_config.xml 文件,文件内容如下:

```xml
<?xml version = "1.0" encoding = "utf-8"?>
<network - security - config >
    < base - config cleartextTrafficPermitted = "true" />
</network - security - config >
```

（2）在 AndroidManifest.xml 文件中的< application >标签下增加以下属性：

```xml
< application
    ...
    android:networkSecurityConfig = "@xml/network_security_config"
/>
```

这样 App 就可以访问 HTTP 的链接了。

第四种方法的原理与第三种方法相同,就是在 AndroidManifest.xml 配置文件的< application >标签中直接插入配置语句,代码如下所示：

```xml
< application
    ...
    android:usesCleartextTraffic = "true"
/>
```

视频讲解

### 9.2.2　HttpURLConnection 接口使用案例

接下来将通过一个输入网址查看图片的例子来讲解 HttpURLConnection 的使用过程。具体过程如下所示。

#### 1. 创建 Chapter9_HttpURLConnection 项目

修改 activity_main.xml 布局界面的代码,整体采用线性垂直布局的方式,放置一个 ImageView 组件、一个 EditText 组件、一个 Button 组件,分别用来显示图片、输入网址、单击显示图片。布局的具体代码如下所示：

```xml
<?xml version = "1.0" encoding = "utf-8"?>
< LinearLayout xmlns:android = "http://schemas.android.com/apk/res/android"
    xmlns:tools = "http://schemas.android.com/tools"
    android:layout_width = "match_parent"
    android:layout_height = "match_parent"
    tools:context = "com.jxust.cn.chapter9_httpurlconnection.MainActivity"
    android:orientation = "vertical">
    < EditText
        android:id = "@ + id/address"
        android:layout_width = "match_parent"
        android:layout_height = "wrap_content"
        android:text = ""
```

```
            http://p1.wmpic.me/article/2017/01/04/1483516503_AQvZcSsM.jpg"/>
<Button
    android:id = "@ + id/get_show"
    android:layout_width = "match_parent"
    android:layout_height = "wrap_content"
    android:text = "获取并显示图片"/>
<ImageView
    android:id = "@ + id/images"
    android:layout_width = "wrap_content"
    android:layout_height = "wrap_content" />
</LinearLayout>
```

**2. 编写 MainActivity 代码**

当界面创建完成以后,需要在 MainActivity 里面编写与界面交互的代码。用于实现请求指定的网络图片,并将获取的图片显示在 ImageView 组件上。具体代码如下所示:

```java
public class MainActivity extends AppCompatActivity {
    private ImageView iv;
    private Button show_btn;
    private EditText path_edit;
    //定义获取到图片和失败的状态码
    protected static final int SUCCESS = 1;
    protected static final int ERROR = 2;
    //创建消息处理器
    private Handler handler = new Handler(){
        public void handleMessage(android.os.Message msg){
            if (msg.what == SUCCESS){
                Bitmap bitmap = (Bitmap)msg.obj;
                iv.setImageBitmap(bitmap);
            }else if (msg.what == ERROR){
Toast.makeText(MainActivity.this,"显示图片错误",Toast.LENGTH_SHORT).show();
            }
        }
    };
    @Override
    protected void onCreate(Bundle savedInstanceState) {
        super.onCreate(savedInstanceState);
        setContentView(R.layout.activity_main);
        init();
    }
    //组件初始化
    private void init(){
        iv = (ImageView)findViewById(R.id.images);
        show_btn = (Button)findViewById(R.id.get_show);
        path_edit = (EditText)findViewById(R.id.address);
        show_btn.setOnClickListener(new View.OnClickListener() {
            @Override
```

```java
public void onClick(View view) {
    //获取输入的网络图片地址
    final String path = path_edit.getText().toString().trim();
    if (TextUtils.isEmpty(path)) {
        Toast.makeText(MainActivity.this,"图片路径不能为空",Toast.LENGTH_SHORT).show();
    } else {
        /* 使用子线程访问网络,因为网络请求耗时,
           4.0以后就不能放在主线程中了 */
        new Thread(){
            private HttpURLConnection conn;
            private Bitmap bitmap;
            public void run(){
                //连接服务器 GET 请求
                try{
                    URL url = new URL(path);
                    //根据URL发送HTTP的请求
                    conn = (HttpURLConnection)url.openConnection();
                    //设置请求的方式
                    conn.setRequestMethod("GET");
                    //设置超时时间
                    conn.setConnectTimeout(5000);
                    //设置请求头 User - Agent 浏览器的版本
                    //得到服务器返回的响应码
                    int state = conn.getResponseCode();
                    Log.v("1111111111",state + "");
                    if(state == 200){
                        //请求网络成功,获取输入流
                        InputStream in = conn.getInputStream();
                        //将流转换为Bitmap对象
                        bitmap = BitmapFactory.decodeStream(in);
                        //告诉消息处理器显示图片
                        Message msg = new Message();
                        msg.what = SUCCESS;
                        msg.obj = bitmap;
                        handler.sendMessage(msg);
                    }else {
                        //请求网络失败,提示用户
                        Message msg = new Message();
                        msg.what = ERROR;
                        handler.sendMessage(msg);
                    }
                } catch (Exception e) {
                    e.printStackTrace();
                    Message msg = new Message();
                    msg.what = ERROR;
                    handler.sendMessage(msg);
                }
            }
        }.start();
    }
}
```

```
        });
    }
}
```

上述代码中的核心部分是先定义一个 URL 对象,然后通过 URL 对象去获取 HttpURLConnection 对象,然后设置请求的方法、超时时间,最后获取到服务器返回的输入流。在网络访问中定义了一个新的线程(new Thread()),网络请求成功后利用 Handle 和 Message 消息处理机制,将访问结果展示在 UI 主线程的 ImageView 中。

**3. 添加访问网络权限**

由于访问网络图片需要请求网络,同时涉及 HTTP 链接的网络访问权限问题,所以需要在清单文件中添加和配置网络权限。具体代码如下:

```
< uses - permission android:name = "android.permission.INTERNET" />
< application
    ...
    android:usesCleartextTraffic = "true"
/>
```

图 9-2 获取并显示图片结果

**4. 运行程序**

此处为了方便,就提前在 EditText 中输入了网络图片的地址,图片地址为 http://p1.wmpic.me/article/2017/01/04/1483516503_AQvZcSsM.jpg,单击"获取并显示图片"按钮,出现的结果如图 9-2 所示。

从图 9-2 可以看出,使用 HttpURLConnection 的 GET 方式获取指定地址的图片,成功地从服务器返回并且显示出来。

### 9.2.3 HttpClient 简介

HttpClient 是 Apache Jakarta Common 下的子项目,用来提供高效的、最新的、功能丰富的支持 HTTP 的客户端编程工具包,并且它支持 HTTP 最新的版本和建议。HttpClient 已经应用在很多的项目中,比如 Apache Jakarta 上很著名的另外两个开源项目 Cactus 和 HTMLUnit 都使用了 HttpClient。

**1. HttpClient 的特性**

(1) 基于标准、纯净的 Java 语言。实现了 HTTP 1.0 和 HTTP 1.1。

(2) 以可扩展的面向对象的结构实现了 HTTP 全部的方法(GET、POST、PUT、DELETE、HEAD、OPTIONS 和 TRACE)。

(3) 支持 HTTPS。

(4) 通过 HTTP 代理建立透明的连接。

(5) 利用 Connect()方法通过 HTTP 代理建立隧道的 HTTPS 连接。

### 2. HttpClient 的使用方法

使用 HttpClient 发送请求、接收响应很简单。具体步骤如下所示：

（1）创建 HttpClient 对象。

（2）创建请求方法的实例，并指定请求 URL。

（3）发送请求参数时，调用 HttpGet、HttpPost 的 setParams(HetpParams params)方法来添加请求参数；对于 HttpPost 对象而言，也可调用 setEntity(HttpEntity entity)方法来设置请求参数。

（4）调用 HttpClient 对象的 execute(HttpUriRequest request)方法发送请求，该方法返回一个 HttpResponse 对象。

（5）调用 HttpResponse 的 getAllHeaders()、getHeaders(String name)等方法可获取服务器的响应头；调用 HttpResponse 的 getEntity()方法可获取 HttpEntity 对象，该对象包装了服务器的响应内容。程序可通过该对象获取服务器的响应内容。

（6）释放连接。无论执行方法是否成功，都必须释放连接。

接下来将介绍使用 HttpClient 访问网络时所需要用到的几个类，具体说明如表 9-2 所示。

表 9-2  HttpClient 常用类说明

| 类 名 称 | 说 明 |
| --- | --- |
| HttpClient | 请求网络的接口 |
| DefaultHttpClient | 实现了 HttpClient 接口的类 |
| HttpGet | GET 请求方法的需要创建的实例 |
| HttpPost | POST 请求方法的需要创建的实例 |
| NameValuePair | 传递参数时的键值对 |
| HttpResponse | 封装了服务器返回的信息类 |
| HttpEntity | 封装了服务返回数据的类 |

接下来将通过具体的例子讲解 HttpClient 的使用。

### 9.2.4　HttpClient 的使用案例

视频讲解

本案例同 9.2.2 节案例相同，都是获取网络图片并显示出来，不同的是本案例是使用 HttpClient 来获取图片。具体操作步骤如下。

**1. 创建 Chapter9_HttpClient 项目**

由于同 9.2.2 节获取图片的本质是一样的，所以布局方式相同，本节就不再对布局代码做详细的介绍，具体可参考 9.2.2 节的布局代码。

**2. 编写 MainActivity 代码**

在 MainActivity 里面编写实现 HttpClient 访问网络图片并在界面显示的逻辑代码，具体代码如下所示：

```
public class MainActivity extends AppCompatActivity {
    private ImageView iv;
    private Button show_btn;
```

```java
private EditText path_edit;
//定义获取到图片和失败的状态码
protected static final int SUCCESS = 1;
protected static final int ERROR = 2;
//创建消息处理器
private Handler handler = new Handler(){
    public void handleMessage(android.os.Message msg){
        if (msg.what == SUCCESS){
            Bitmap bitmap = (Bitmap)msg.obj;
            iv.setImageBitmap(bitmap);
        }else if (msg.what == ERROR){
Toast.makeText(MainActivity.this,"显示图片错误",Toast.LENGTH_SHORT).show();
        }
    }
};
@Override
protected void onCreate(Bundle savedInstanceState) {
    super.onCreate(savedInstanceState);
    setContentView(R.layout.activity_main);
    init();
}
//组件初始化
private void init(){
    iv = (ImageView)findViewById(R.id.images);
    show_btn = (Button)findViewById(R.id.get_show);
    path_edit = (EditText)findViewById(R.id.address);
    show_btn.setOnClickListener(new View.OnClickListener() {
        @Override
        public void onClick(View view) {
            //获取输入的网络图片地址
            final String path = path_edit.getText().toString().trim();
            if(TextUtils.isEmpty(path)){
Toast.makeText(MainActivity.this,"图片路径不能为空",Toast.LENGTH_SHORT).show();
            }else {
//使用子线程访问网络,因为网络请求耗时,在 Android 4.0 以后就不能放在主线程中了
                new Thread(){
                    private HttpURLConnection conn;
                    private Bitmap bitmap;
                    public void run() {
                        //使用 HttpClient 获取图片
                        getImageByHttpClient(path);
                    }
                }.start();
            }
        }
    });
```

```java
        }
        //获取图片方法
        private void getImageByHttpClient(String path) {
            //获取 HttpClient 对象
            HttpClient Client = new DefaultHttpClient();
            HttpGet get = new HttpGet(path);
            try{
                //获取返回的 HttpResponse 对象
                HttpResponse response = Client.execute(get);
                //查看状态码是否为 200
                if(response.getStatusLine().getStatusCode() == 200){
                    //请求成功,获取 HttpEntity 对象
                    HttpEntity entity = response.getEntity();
                    //获取输入流
                    InputStream in = entity.getContent();
                    //获取 Bitmap 对象
                    Bitmap bitmap = BitmapFactory.decodeStream(in);
                    //通知消息处理器显示图片
                    Message msg = new Message();
                    msg.what = SUCCESS;
                    msg.obj = bitmap;
                    handler.sendMessage(msg);
                }else {
                    Message msg = new Message();
                    msg.what = ERROR;
                    handler.sendMessage(msg);
                }
            }catch (Exception e){
                Message msg = new Message();
                msg.what = ERROR;
                handler.sendMessage(msg);
            }
        }
    }
```

上述代码中,只有在 getImageByHttpClient(String path)方法里的代码与使用 HttpURLConnection 不同。此处方法中采用的是 HttpGet()方法请求获取网络图片资源,访问成功后会返回 200 的状态码,然后使用 HttpResponse 的 getEntity()方法获得 HttpEntity 对象,然后通过 HttpEntity 对象的 getContent()方法得到输入流,最后转换为 Bitmap 对象显示出来。

需要注意的是,Google 公司从 Android 6.0(API 23)以后,就不建议再用 HttpClient 了,取而代之的是 HttpUrlConnection,所以要使用 HttpClient 就需要在 build.gradle 加入如下语句:

```
android {
    useLibrary 'org.apache.http.legacy'
}
```

## 3. 添加权限

本案例同样需要访问网络,所以需要添加网络权限,同时还要添加对 HttpClient 兼容性的设置,代码如下:

```
< uses - permission android:name = "android.permission.INTERNET" />
< application
    ...
    android:usesCleartextTraffic = "true"
/>
< uses - library android:name = "org.apache.http.legacy" android:required = "false" />
```

## 4. 运行程序

代码编写完以后,运行程序,单击"获取并显示图片"按钮,出现的结果如图 9-3 所示。

图 9-3 获取并显示图片结果

# 9.3 Socket 通信

Android 应用程序与服务器通信的方式主要有两种:一种是 HTTP 通信;另一种是 Socket 通信。HTTP 连接使用的是请求-响应方式,即在请求时才建立连接。而 Socket 通信则是在双方建立连接后直接进行数据传输。它在连接时可实现信息的主动推送,而不用每次都等客户端先向服务器发送请求。

## 9.3.1 Socket 通信原理

Socket 通常称为"套接字",用于描述 IP 地址和端口,是一个通信链的句柄。应用程序通常通过套接字向网络发出请求或者应答网络请求,它支持 TCP/IP 的网络通信的基本单元。它是网络通信过程中端点的抽象表示,包含进行网络通信的 5 种必需信息:连接使用的协议、本地主机的 IP 地址、本地进程的协议端口、远程主机的 IP 地址、远程进程的协议端口。

### 1. 创建 Socket

连接 Socket 连接至少需要两个套接字:一个运行于客户端;另一个运行于服务端。它们都已经封装成类,常用的构造方法如下所示:

(1) Socket(InetAddress address,int port)。
(2) Socket(InetAddress address,int port,boolean stream)。
(3) Socket(String host,int port,Boolean stream)。
(4) ServerSocket(int port)。
(5) ServerSocket(int port,int backlog)。
(6) ServerSocket(int port,int backlog,InetAddress bindAddr)。

上述参数说明如下所示:

Address——双向连接中另一方的 IP 地址。
host——另一方的主机名。
port——另一方的端口号。
stream——指明 Socket 是流还是数据报 Socket。
bindAddr——本地机器的地址。

创建 Socket 的代码如下所示:

```
Socket socket = new Socket("192.168.56.1", "35434");
ServerSocket server = new ServerSocket("35434");
```

需要注意的是,在选择端口时每一个端口对应一个服务。0~1023 的端口号为系统所保留,所以在选择端口号时最好选择一个大于 1023 的数,如上面的 35434,以防止发生冲突。在创建 Socket 时,需要捕获或抛出异常。

### 2. 输入(输出)流

Socket 提供了 getInputStream()和 getOutputStream()来得到对应的输入或者输出流进行读写操作,这两个方法分别返回 InputStream 和 OutputStream 类对象。为了便于读写数据,可以在返回输入、输出流对象上建立过滤流。对于文本方式流对象,可以采用 InputStreamReader()、OutputStreamWriter()和 PrintWriter()处理,具体代码如下所示:

```
PrintStream ps = new PrintStream(new BufferedOutputStream(Socket.getOutputStream()));
//设置过滤流
DataInputStream is = new DataInputStream(socket.getInputStream());
PrintWriter pwriter = new PrintWriter(socket.getOutStream(),true);
BufferedReader breader = new BufferedReader(new InputStreamReader(Socket.getInputStream));
```

### 3. 关闭 Socket 流

在 Socket 使用完毕后需要将其关闭,以释放资源。需要注意的是,在关闭 Socket 之前,需要将与 Socket 相关的输入输出流先关闭。具体代码如下所示:

```
ps.close();              //输出流先关闭
is.close();              //输入流其次关闭
socket.close();          //Socket 最后关闭
```

## 9.3.2　Socket 通信案例

视频讲解

本案例是实现编写客户端负责发送内容,服务端用来接收内容的程序,具体步骤如下。

### 1. 编写服务端程序

该程序是负责接收数据,需要单独编译运行。具体代码如下所示:

```java
public class test_Socket implements Runnable {
    public static final String Server_ip = "127.0.0.1";
    public static final int Server_port = 2000;
    @Override
    public void run() {
        System.out.println("S:Connectioning...");
        try {
            ServerSocket serverSocket = new ServerSocket(Server_port);
            while (true){
                Socket client = serverSocket.accept();
                System.out.println("S:Receing...");
                try {
                    BufferedReader breader = new BufferedReader(new
                            InputStreamReader(client.getInputStream()));
                    String str = breader.readLine();
                    System.out.println("S:Received:" + str);
                }catch (Exception e){
                    System.out.print("S:Error");
                    e.printStackTrace();
                }finally {
                    client.close();
                    System.out.println("S:Done");
                }
            }
        } catch (IOException e) {
            e.printStackTrace();
        }
    }
    public static void main(String[] args){
        Thread thread = new Thread(new test_Socket());
        thread.start();
    }
}
```

上述代码中设置了服务器端口为2000,然后通过 accept()方法使服务器开始监听客户端的连接,然后通过 BufferReader 对象来接收输入流。最后关闭 Socket 和流。

## 2. 编写客户端布局文件

客户端布局界面包含一个 EditText 组件和一个按钮，按钮负责把输入的内容发送到服务器端。activity_main.xml 的具体代码如下所示：

```xml
<?xml version = "1.0" encoding = "utf-8"?>
<LinearLayout>
    <EditText
        android:id = "@+id/mes"
        android:layout_width = "match_parent"
        android:layout_height = "wrap_content"
        android:hint = "请输入要发送的消息"/>
    <Button
        android:id = "@+id/send"
        android:layout_width = "match_parent"
        android:layout_height = "wrap_content"
        android:text = "发送消息"/>
</LinearLayout>
```

## 3. 编写 MainActivity 代码

MainActivity 负责客户端的实现，在按钮事件中通过"socket＝new Socket(ip,port)"请求连接服务器，并通过 BufferedWriter() 发送消息，具体代码如下所示：

```java
public class MainActivity extends AppCompatActivity {
    private EditText mes;
    private Button send_btn;
    private String ip = "172.16.39.192";
    private int port = 2000;
    @Override
    protected void onCreate(Bundle savedInstanceState) {
        super.onCreate(savedInstanceState);
        setContentView(R.layout.activity_main);
        if (SDK_INT > 8)
        {
            StrictMode.ThreadPolicy policy = new StrictMode.ThreadPolicy.Builder()
                    .permitAll().build();
            StrictMode.setThreadPolicy(policy);
        }
        init();
    }
    //组件初始化方法
    private void init(){
        mes = (EditText)findViewById(R.id.mes);
        send_btn = (Button)findViewById(R.id.send);
        send_btn.setOnClickListener(new View.OnClickListener() {
            @Override
            public void onClick(View view) {
```

```java
            try {
                String string = mes.getText().toString();
                if (!TextUtils.isEmpty(string)){
                    SendMes(ip,port,string);
                }else {
                    Toast.makeText(MainActivity.this,"请先输入内容",Toast.LENGTH_SHORT).show();
                    mes.requestFocus();
                }
            }catch (Exception e){
                e.printStackTrace();
            }
        }
    });
}
private void SendMes(String ip, int port, String mes) throws UnknownHostException, IOException{
    try {
        Socket socket = null;
        socket = new Socket(ip,port);
        BufferedWriter writer = new BufferedWriter(new
                OutputStreamWriter(socket.getOutputStream()));
        writer.write(mes);
        writer.flush();
        writer.close();
        socket.close();
    }catch (UnknownHostException e ){
        e.printStackTrace();
    }catch (IOException e){
        e.printStackTrace();
    }
}
}
```

### 4. 运行

运行程序,输入要发送的消息,然后单击"发送消息"按钮,出现的结果如图 9-4 和图 9-5 所示。

图 9-4　客户端界面

图 9-5　服务端结果

从上面的运行结果可以看出,客户端与服务端连接成功并且发送了消息。这就是 Socket 通信的具体步骤,开发者可以使用者开发一些更为复杂的应用。

## 9.4 数据提交方式

HTTP 1.1 规定的 HTTP 请求方法有 OPTIONS、GET、HEAD、POST、PUT、DELETE、TRACE、CONNECT。其中 POST 一般用来向服务端提交数据,GET 请求一般用来表示客户端请求一个 Uri,服务器返回客户端请求的 Uri。接下来将详细讲解 GET 和 POST 两种请求方式的差异。

### 9.4.1 GET 方式提交数据

GET 的本质是从服务器获取数据,效率比 POST 高。GET 请求能够被缓存,在 HTTP 的定义中,没有对 GET 请求的数据大小限制,不过因为浏览器不同,一般限制在 2~8KB。

GET 发送请求时,URL 中除了资源路径以外,所有的参数(查询字符串)也封装在 URL 中,并且记录在服务器的访问日志中,所以不要传递一些例如身份证信息、密码等敏感信息。

**1. 参数格式**

在资源路径末尾添加"?"表示追加参数。每一个变量及值按照"变量名=变量值"方式设定,不能包含空格或者中文。

多个参数使用"&"连接。

**注意**:URL 字符串中如果包含空格或者中文,需要添加百分号转义。

**2. GET 方式提交数据代码**

接下来将通过一段代码讲解如何使用 HttpURLConnection 的 GET 方式提交数据。具体代码如下所示:

```
//使用子线程访问网络,因为网络请求耗时,在4.0以后就不能放在主线程中了
new Thread(){
    private HttpURLConnection conn;
    private Bitmap bitmap;
    public void run(){
        //连接服务器GET请求
        try{
            URL url = new URL(path);
            //根据URL发送HTTP的请求
            conn = (HttpURLConnection)url.openConnection();
            //设置请求的方式
            conn.setRequestMethod("GET");
            //设置超时时间
            conn.setConnectTimeout(5000);
            //设置请求头 User-Agent 浏览器的版本
            //得到服务器返回的响应码
            int state = conn.getResponseCode();
            Log.v("HttpURL",state + "");
            if(state == 200){
```

```
                    //请求网络成功,获取输入流
                
            }else {
                //请求网络失败,提示用户
            
            }
        } catch (Exception e) {
            e.printStackTrace();
        
        }
    }
}.start();
```

以上就是如何使用 HttpURLConnection 的 GET 请求去提交图片网址到服务器并且获取网络图片显示在应用程序的主界面上的主要代码。

## 9.4.2　POST 方式提交数据

视频讲解

POST 的本质是向服务器发送数据,也可以获得服务器处理之后的结果,效率不如 GET。

POST 请求不能被缓存,POST 提交数据比较大,其大小靠服务器的设定值限制,PHP 通常限定为 2MB。

POST 发送请求时,URL 中只有资源路径,但不包含参数,服务器日志不会记录参数,相对更安全。

参数被包装成二进制的数据形式,格式与 GET 基本一致,只是不包含"?"。

**注意**：所有涉及用户隐私的数据(密码、银行卡号)一定记住使用 POST 方式传递,浏览器可以监视 POST 请求,但是不容易捕捉到。

接下来将通过一段代码讲解如何使用 HttpClient 的 POST 方式提交数据。具体代码如下所示：

```
/* 利用 HttpClient 进行 POST 请求的工具类
 */
public class HttpClientUtil {
    public String doPost(String url, Map<String, String> map, String charset){
        HttpClient httpClient = null;
        HttpPost httpPost = null;
        String result = null;
        try{
            httpClient = new SSLClient();
            httpPost = new HttpPost(url);
            //设置参数
            List<NameValuePair> list = new ArrayList<NameValuePair>();
            Iterator iterator = map.entrySet().iterator();
            while(iterator.hasNext()){
                Entry<String, String> elem = (Entry<String, String>) iterator.next();
```

```
            list.add(new BasicNameValuePair(elem.getKey(), elem.getValue()));
        }
        if(list.size() > 0){
            UrlEncodedFormEntity entity = new UrlEncodedFormEntity(list, charset);
            httpPost.setEntity(entity);
        }
        HttpResponse response = httpClient.execute(httpPost);
        if(response != null){
            HttpEntity resEntity = response.getEntity();
            if(resEntity != null){
                result = EntityUtils.toString(resEntity, charset);
            }
        }
    }catch(Exception ex){
        ex.printStackTrace();
    }
    return result;
}
```

以上就是使用 HttpClient 发送 POST 请求的部分代码,使用 POST 请求传递参数的代码如上述代码中的加粗代码所示。关于使用 HttpClient 发送 POST 请求,在后续的章节中会通过与 PHP 结合的方式来更详细地讲解如何使用。

需要注意的是,在日常的开发中,手机端与服务器进行数据交互时,可能会出现中文乱码的问题,所以在发送请求时需要设置编码。

## 9.5 OkHttp 的使用

视频讲解

### 9.5.1 OkHttp 的 GET 和 POST 请求

OkHttp 是一款优秀的 HTTP 框架,它支持 GET 请求和 POST 请求,支持基于 HTTP 的文件上传和下载、支持加载图片、支持下载文件透明的 GZIP 压缩、支持响应缓存避免重复的网络请求、支持使用连接池降低响应延迟问题。

**1. 同步 GET 请求**

同步 GET 请求首先需要初始化 OkHttpClient 对象,然后构造 Request 对象,接着使用前两部的对象来构建 Call 对象,最后通过 call.execute()来提交请求。这种方式会阻碍调用线程,所以在 Android 中应放在子线程中执行,否则有可能引起 ANR 异常,Android 不允许在主线程中进行耗时性操作。GET 请求的具体代码如下所示:

```
String url = "https://github.com/square/okhttp";
OkHttpClient okHttpClient = new OkHttpClient();
final Request request = new Request.Builder().url(url).build();
```

```
final Call call = okHttpClient.newCall(request);
new Thread(new Runnable() {
    @Override
    public void run() {
        try {
            Response response = call.execute();
        } catch (IOException e) { e.printStackTrace();}
    }
}).start();
```

### 2. 异步 GET 请求

异步 GET 请求与同步 GET 请求步骤基本一致，但是通过调用 call.enqueue(new Callback)方法来提交异步请求，这种方式发出的异步请求会被加入 Dispatcher 中的 runningAsyncCalls 双端队列中通过线程池来执行。异步 GET 请求具体内容如下：

```
String url = " https://github.com/square/okhttp ";
OkHttpClient okHttpClient = new OkHttpClient();
final Request request = new Request.Builder().url(url).get().build();
Call call = okHttpClient.newCall(request);
call.enqueue(new Callback() {
    @Override
    public void onFailure(Call call, IOException e) {
    }
    @Override
    public void onResponse(Call call, Response response) throws IOException {
    }
});
```

### 3. 异步 POST 请求

同步 POST 请求与 GET 类似，这里就不再讲述，只讲述异步 POST 请求的方式。首先看一段异步 POST 请求的代码：

```
private void postDataWithParame() {
    OkHttpClient client = new OkHttpClient();                    //创建 OkHttpClient 对象
    FormBody.Builder formBody = new FormBody.Builder();          //创建表单请求体
    formBody.add("username","zhangsan");                         //传递键值对参数
    Request request = new Request.Builder()                      //创建 Request 对象
        .url("https://github.com/square/okhttp ").post(formBody.build()).build();
    client.newCall(request).enqueue(new Callback() {...});
}
```

从上述代码可以看出，POST 请求并没有设置请求方式为 POST，GET 请求也同样没有设置，那是通过什么来区分呢？从上述加粗部分的代码中可以看到，Request.Builder 对象创建最初默认

是 GET 请求,因此在 GET 请求中不需要设置请求方式,当调用 POST()方法时,请求方式修改为 POST。

上述代码中传递的参数是通过创建表单请求体对象,然后把表单请求体对象作为 post()方法的参数。POST 请求传递参数的方法还有很多种,但都是通过 post()方法传递的,Request.Builder 类的 post()方法的声明如下所示:

```
public Builder post(RequestBody body)
```

从上面方法的声明可以看出,post()方法接收的参数是 RequestBody 对象,所以只要是 RequestBody 类以及子类对象都可以当作参数进行传递。上述使用的 FormBody 就是一个子类对象。

RequestBody 是抽象类,故不能直接使用,但是它有静态方法 create(),使用这个方法可以得到 RequestBody 对象。使用这种方式可以上传 JSON 对象或 File 对象。上传 JSON 对象具体的内容如下:

```
OkHttpClient client = new OkHttpClient();              //创建 OkHttpClient 对象
//数据类型为 JSON 格式
MediaType JSON = MediaType.parse("application/json; charset=utf-8");
String jsonStr = "{\"username\":\"赵二\",\"nickname\":\"赵二\"}";   //JSON 数据
RequestBody body = RequestBody.create(JSON,josnStr);
Request request = new Request.Builder()
    .url("http://www.baidu.com").post(body).build();
client.newCall(request).enqueue(new Callback() {...});
```

上传 File 对象的具体方式如下所示:

```
OkHttpClient client = new OkHttpClient();              //创建 OkHttpClient 对象
MediaType fileType = MediaType.parse("File/*");        //数据类型为 JSON 格式
File file = new File("path");                          //File 对象
RequestBody body = RequestBody.create(fileType,file);
Request request = new Request.Builder()
    .url("https://github.com/square/okhttp").post(body).build();
client.newCall(request).enqueue(new Callback(){...});
```

其他的方式如流的方式、分块提交的方式等这里就不再介绍,读者可以自行学习调试代码。

### 9.5.2　在 Android 中使用 OkHttp

视频讲解

要在 Android 中使用 OkHttp,首先需要在 build.gradle 中添加第三方的库,GitHub 上显示目前最新版本是 4.2.2,所以在 build.gradle 中添加 dependencies 依赖。具体如下所示:

```
implementation "com.squareup.okhttp3:okhttp:4.2.2"
```

添加完依赖后,需要在 AndroidManifest.xml 中添加网络权限,9.2.1 节已经介绍过,这里不再

赘述。本案例使用最简单的 Textview 去显示 GET 或者 POST 请求获取的信息。MainActivity 的具体内容如下所示：

```java
public class MainActivity extends AppCompatActivity {
    private TextView textView;
    @Override
    protected void onCreate(Bundle savedInstanceState) {
        super.onCreate(savedInstanceState);
        setContentView(R.layout.activity_main);
        textView = findViewById(R.id.texts);
        postaAsync();
    }
    //GET 方式(省略)
    private void postaAsync() {
        MediaType mediaType = MediaType.parse("text/x-markdown; charset=utf-8");
        String requestBody = "I am Jdqm.";
        Request request = new Request.Builder()
                .url("https://api.github.com/markdown/raw")
                .post(RequestBody.create(mediaType, requestBody))
                .build();
        OkHttpClient okHttpClient = new OkHttpClient();
        okHttpClient.newCall(request).enqueue(new Callback() {
            @Override
            public void onFailure(Call call, IOException e) {
            }
            @Override
            public void onResponse(Call call, Response response) throws IOException {
                Headers headers = response.headers();
                showResponse(headers.toString());
            }
        });
    }
    private void showResponse(final String response) {
        runOnUiThread(new Runnable() {
            @Override
            public void run() {
                textView.setText(response);
            }
        });
    }
}
```

以上代码主要展示了 OkHttp 的 GET 和 POST 请求方式的应用，实现的效果如图 9-6 所示。

在第 12 章中，将会结合本节的知识去完善"橙一色"App 项目，通过请求本地服务器端的接口去获取本地数据库中的数据等，进而实现"橙一色"水果数据的实时变更。

图 9-6　OkHttp 示例图

## 9.6　本章小结

本章主要讲解了 Android 系统的网络通信编程。首先介绍 HTTP，讲解使用 HttpURLConnection、HttpClient 访问网络资源的方式。接着讲解 Android 中的 Socket 通信，通过一个客户端与服务器端通信的案例讲解 Socket 的使用过程。然后介绍 HTTP 中常用的两种数据提交方式：GET、POST，通过示例代码说明了这两种请求方式的使用差异。最后介绍 OKHttp 的使用。由于在日常的应用程序开发中涉及网络通信部分的较多，本章内容需要熟练掌握。

## 9.7　课后习题

1. 简述 HttpClient 和 HttpURLConnection 访问网络的步骤。
2. 简述 Socket 通信的步骤。
3. 简述 GET、POST 请求方式的差别。
4. 使用 HttpClient 访问一个主页，并把获取到的内容显示出来。
5. 使用 OkHttp 访问一个主页，并把获取到的内容显示出来。

# 第 10 章

# Android+PHP 开发

**学习目标**
- 掌握 PHP+MySQL 的结合使用。
- 掌握 PHP 对数据库的基本操作。
- 掌握 Android+PHP+MySQL 的开发。

在第 9 章讲解了 Android 网络通信编程的知识,由于在网络编程中,应用程序需要向服务器提交数据,所以涉及后台服务器的开发。对于 Android 的开发,通常使用 PHP 作为后台,连接数据库和应用程序之间的交互。接下来将详细介绍 Android+PHP 的开发。

## 10.1 PHP 介绍

PHP(Hypertext Preprocessor,超文本预处理语言)是一种动态网页开发语言,其语法吸收了 C 语言、Java 和 Perl 的特点,便于学习,使用广泛,主要适用于 Web 开发领域。

**1. PHP 的特性**

(1) PHP 独特的语法混合了 C、Java、Perl 以及 PHP 自身创新的语法。
(2) PHP 可以比 CGI 或者 Perl 更快速地执行动态网页。
(3) 所有比较受欢迎的数据库以及操作系统基本上都可以使用 PHP 开发。
(4) 最重要的是 PHP 可以用 C、C++语言进行程序的扩展。

**2. PHP 的优点**

(1) 免费性:和其他技术相比,PHP 本身免费且是开源代码。
(2) 快捷性:程序开发速度快,运行效率高,技术本身比较容易学习掌握。
(3) 跨平台性强。
(4) 效率高:PHP 占用相当少的系统资源。
(5) 开放源代码:所有的 PHP 源代码事实上都可以得到。

### 3. PHP 开发环境搭建

要进行 PHP 的开发,首先需要搭建 PHP 的开发环境,本节所讲的是针对在 Windows 平台上的 PHP 开发环境搭建的。

PHP 服务器组件非常多,有 WampServer、XAMPP、AppServ、phpStudy、phpNow 等。WampServer 目前在 Windows 平台上使用最广泛,操作也非常简单,WampServer 内部还集成了 phpMyAdmin 数据库管理工具。接下来将详细讲解如何使用 WampServer 搭建 PHP 的开发环境。具体步骤如下。

(1) 下载 PHP 开发环境所需要组件:WampServer 软件,打开浏览器,输入 http://www.wampserver.com/en/#download-wrapper,如图 10-1 所示。

图 10-1 下载 WampServer 软件

(2) 打开下载的 WampServer 安装包(这里用的 WampServer 3.2.0 版本),选择安装路径,然后单击 Next 按钮一直按照提示安装即可。路径选择的操作如图 10-2 所示。

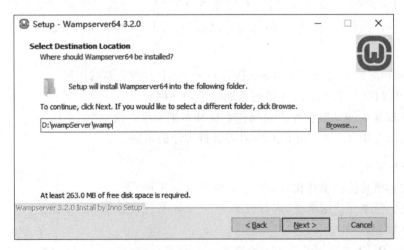

图 10-2 安装 WampServer

（3）安装完 WampServer 以后，单击桌面上的图标，启动 WampServer 软件，然后在桌面右下角会出现图标 。单击该图标，显示当前 WampServer 以及 phpMyAdmin、Apache、PHP、MySQL 的版本号，如图 10-3 所示。

（4）单击图 10-3 中的 MySQL 图标，显示当前 MySQL 的端口号为 3308，如图 10-4 所示。通常情况下，MySQL 默认端口号为 3306，如果在安装 WampServer 之前用户已单独安装了 MySQL，3306 端口号被占用，则此时 WampServer 集成的 MySQL 就会在安装时自动改变端口号。启动 WampServer 服务时要注意端口号，如果是默认的 3306，可以省略不写，如果是新的端口号，则在数据库配置中要写上端口号。

图 10-3　WampServer 的运行界面

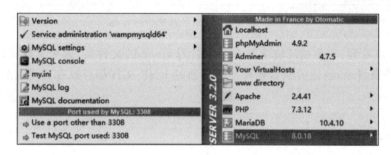

图 10-4　WampServer 中的 MySQL

（5）接下来测试 PHP 的开发环境是否搭建成功。打开浏览器，在地址栏输入 localhost，然后按 Enter 键，出现的结果如图 10-5 所示。

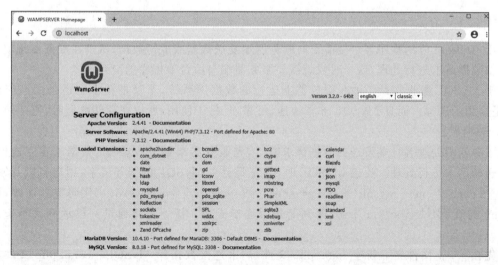

图 10-5　测试界面

若出现上述界面，则代表 PHP 开发环境已经搭建成功，接下来就可以进行 PHP 的开发了。

### 4. PHP 创建的项目位置

在 WampServer 安装成功以后,打开 WampServer 所在的文件夹,然后可以看到文件夹的结构如图 10-6 所示。

图 10-6 PHP 开发环境文件夹结构

从图 10-6 可以看到,安装的环境下存在一个名为 www 的文件夹,这个文件夹就是开发者所创建的 PHP 项目的根目录,创建的项目或者一个单独的 PHP 文件都存放在这个文件夹下,然后才能运行。

## 10.2 PHP+MySQL 简介

视频讲解

10.1 节讲解了 PHP 的特性、优点以及开发环境的搭建,本节将讲解 PHP 和 MySQL 的连接、使用 PHP 创建数据库,使用 PHP 创建数据库表以及对数据表的操作等。

### 10.2.1 PHP 连接 MySQL

由于 Android 应用程序客户端需要向服务器端发送数据或取得数据,所以在服务器端就需要对客户端的数据请求进行处理,这个时候就需要服务器端与后台数据库的交互。

由于 MySQL 数据库是一种在服务器上运行的数据库系统,不管是在小型还是大型应用程序中,都是理想的选择。而且 MySQL 非常快速、可靠,且易于使用,所以在存储数据时,通常使用的是 MySQL 数据库。

在服务器端处理客户端发送的数据请求时,首先服务器端要与 MySQL 数据库建立连接,然后再对发送的请求进行处理。连接 MySQL 时,首先要定义 MySQL 的服务名字、用户名以及密码,然后使用"$conn = new mysqli($servername, $username, $password);"代码建立连接,可以根据 $conn 的值是不是 connect_error 来判断是否连接成功。接下来将通过一段示例代码来说明如何操作。

首先在 www 文件夹下新建 conn.php 文件,用来连接数据库 MySQL。conn.php 文件的具体内容如下所示:

```
<?php
    header("Content-type: text/html; charset=utf-8");
```

```php
$ servername = "localhost:3308";            //WampServer 中 MySQL 的端口号
$ username = "root";
$ password = "";
$ conn = new mysqli( $ servername, $ username, $ password);
if ( $ conn->connect_error) {
   die("连接失败: " . $ conn->connect_error);   }
echo "连接成功";
?>
```

完成以后,在浏览器地址栏中输入 http://localhost/conn.php 查看结果。出现的结果如图10-7所示。

需要注意的是,在使用完数据库以后,需要关闭数据库的连接。对于上述建立连接的方式来说,关闭连接的代码如下所示:

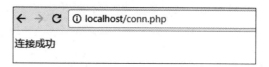

图 10-7　使用 PHP 连接 MySQL 成功的结果

```php
$ conn->close();
```

以上就是使用 MySQLi(面向对象)的方式建立连接,也可以使用 MySQLi(面向过程)的方式或者 PDO 方式建立连接。需要注意的是,使用不同的方式连接时,判断的条件不太一样,关闭连接的方式也不太一样。

### 10.2.2　PHP 创建数据库

在学习了 PHP 与 MySQL 建立连接以后,就可以使用 MySQL 来创建数据库了。PHP 创建数据库的方式有两种:一种是手动创建;另一种是使用代码来创建。接下来将详细讲解这两种创建数据库的方式。

**1. 手动创建数据库**

手动创建数据库的方式比较简单,首先在浏览器地址栏中输入 localhost,然后单击主界面上的 Tools 目录下的 phpmyadmin。操作如图 10-8 所示。

图 10-8　打开 phpMyAdmin

单击后,出现 phpMyAdmin 登录界面,输入用户名 root,密码保持默认,在"选择服务器"下拉列表框中选择 MySQL,如图 10-9 所示。

单击"执行"按钮,出现数据库界面,如图 10-10 所示。

在图 10-10 中,单击左侧的"新建",出现右侧的操作界面,输入数据库名称,默认的 utf8mb4_

图 10-9　使用 phpMyAdmin 登录 MySQL

图 10-10　创建数据库的操作

0900_ai_ci 表示排序规则(也可选择 utf8_general_ci),单击"创建"按钮,即可创建数据库。

以上就是手动创建数据库的方式,接下来将讲解通过代码创建数据库的方式。

### 2. 通过代码创建数据库

通过代码创建数据库时,首先需要连接 MySQL 数据库,然后再执行创建数据库的语句。具体代码如下所示:

```
<?php
    header("Content-type: text/html; charset=utf-8");
```

```php
    $ servername = "localhost:3308";            //WampServer 中 MySQL 的端口号
    $ username = "root";
    $ password = "";
    $ conn = new mysqli( $ servername, $ username, $ password);
    if ( $ conn -> connect_error) {
        die("连接失败:" . $ conn -> connect_error);}
    $ sql = "CREATE DATABASE user";
    if ( $ conn -> query( $ sql) === TRUE) {
        echo "数据库创建成功";} else {
        echo "Error creating database: " . $ conn -> error;
    }
    $ conn -> close();
?>
```

以上就是通过 PHP 创建 user 数据库的详细代码,命名为 create_db.php 文件,保存到 WampServer 安装路径的 www 目录下,运行该文件,出现的结果如图 10-11 所示。

打开数据库手动操作的界面,可以看到 user 数据库已经创建成功了。界面如图 10-12 所示。

图 10-11 代码运行结果

图 10-12 数据库操作界面

## 10.2.3 PHP 创建数据表

在学习完创建数据库以后,接下来创建数据库表。创建数据库表的操作也有两种:手动创建和通过代码创建。手动创建的方式与手动创建数据库的方式一致,此处不再赘述。接下来讲解通过代码如何创建数据表。

通过代码创建数据表时,首先要与已有的数据库建立连接,如上面代码创建的 user 数据库,与数据库连接以后,再执行创建数据表语句即可创建。

新建一个 create_user_login.php 文件,具体代码如下所示:

```php
<?php
//省略数据库连接信息
if ( $ conn -> connect_error) {
    die("连接失败:" . $ conn -> connect_error);
}
$ sql = "CREATE TABLE user_mes (
id INT(6) AUTO_INCREMENT PRIMARY KEY,
name VARCHAR(30),
```

```
    sex VARCHAR(30),
    email VARCHAR(50)
    )";
    if ( $ conn -> query( $ sql ) === TRUE) {
        echo "user_mes 数据表创建成功";
    } else {
        echo "创建数据表错误: " . $ conn -> error;
    }
    $ conn -> close();
?>
```

运行 create_user_login.php,运行结果如图 10-13 所示。

打开手动操作数据库的界面,然后单击 user 数据库,发现 user_mes 数据库表已经创建成功。界面如图 10-14 所示。

图 10-13　代码运行结果

图 10-14　user 数据库视图

视频讲解

### 10.2.4　PHP 对数据库表的基本操作

PHP 对数据表的基本操作即是数据的增加、删除、查询和修改,连接到数据库以后,执行这些基本操作语句,即可对表的数据进行操作。接下来将针对 PHP 对数据库表添加数据以及查询数据的操作进行详细讲解。

#### 1. 使用 PHP 添加数据

在使用 PHP 添加数据时,可以手动添加,也可以通过代码添加,这里就不再讲述手动添加的方式,接下来将详细地讲解如何通过代码添加数据。

新建一个 insert_user_mes.php 文件,执行该文件代码时首先要连接数据,然后执行插入语句。该文件的具体代码如下所示:

```
<?php
    //省略数据库连接信息及判断
    $ sql = "INSERT INTO user_mes (name, sex, email)
    VALUES ('liming', 'male', '1233333@qq.com')";
    if ( $ conn -> query( $ sql ) === TRUE) {
        echo "新记录插入成功";} else {
        echo "Error: " . $ sql . "<br>" . $ conn -> error;
    }
    $ conn -> close();
?>
```

执行上述代码以后,出现的结果如图 10-15 所示。

打开手动操作数据库的界面,查看 user_mes 表的数据,发现数据已经添加成功了。该表的数据视图如图 10-16 所示。

图 10-15　插入数据结果

图 10-16　user_mes 表数据视图

### 2. 使用 PHP 读取数据

使用 PHP 从数据库中读取数据时,首先需要先连接数据库,然后再读取数据。接下来将通过具体的代码来讲解如何使用 PHP 读取 MySQL 数据库中的数据。

新建一个 query_user_mes.php 文件。该文件的具体代码如下所示:

```php
<?php
//省略数据库连接信息及判断
$sql = "SELECT name, sex, email FROM user_mes";
$result = $conn->query($sql);
if ($result->num_rows > 0) {
    while($row = $result->fetch_assoc()) {
        echo "name:". $row['name'].", sex:". $row['sex'].", email:". $row['email'];
    }
} else {   echo "0 结果";   }
$conn->close();?>
```

以上代码执行的结果如图 10-17 所示。

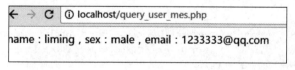

图 10-17　查询数据结果

从图 10-17 可以看出,查询的数据与存入的数据一致,表示查询成功。

## 10.3 PHP+Android 简介

### 10.3.1 Android 与 PHP 结合

在第9章中学习了如何使用 HttpClient 向服务器发送请求,本节就将使用 PHP 作为后台服务器处理请求,在 Android 前端向服务器发送请求。

Android+PHP+MySQL 的处理过程如图 10-18 所示。

图 10-18 前后台交互原理示意图

从图 10-18 可以看出,当用户单击应用程序界面上的某个组件时,发送网络请求到 PHP 文件,也就是服务器端。如果用户需要对数据库操作,此时服务器端就会连接数据库,然后操作数据库,操作完成以后会返回一个结果集到客户端,给予用户想得到的信息。

**1. 客户端网络请求发送类**

接下来将通过代码讲解如何在 Android 前端向 PHP 后台服务器发送请求。首先新建一个文件 Http_Conn.java,该类使用 HttpClient 发送网络请求。具体代码如下所示:

```java
public class Http_Conn {
    //连接的方法
    public boolean gotoConn(String phonenum, String password, String connectUrl) {
        String result = "";                    //用来取得返回的 String
        boolean isLoginSucceed = false;
        HttpClient httpClient = new DefaultHttpClient();
        //发送 POST 请求
        HttpPost httpRequest = new HttpPost(connectUrl);
```

```java
            //POST 运作传送变数必须用 NameValuePair[]阵列存储
            List< NameValuePair > params = new ArrayList< NameValuePair >();
            //BasicNameValuePair 存储键值对的类
            params.add(new BasicNameValuePair("phone", phonenum));
            params.add(new BasicNameValuePair("paswd", password));
            try {
                //发出 HTTP 请求转为带参数的 HTTP 网络地址
                httpRequest.setEntity(new UrlEncodedFormEntity(params,"utf-8"));
                //取得 HTTP Response
                HttpResponse httpResponse = httpClient.execute(httpRequest);
                result = EntityUtils.toString(httpResponse.getEntity());
            } catch (Exception e) {
                e.printStackTrace();
            }
            if (result.equals("success")) {
                isLoginSucceed = true;
            }
            return isLoginSucceed;
        }
    }
```

以上代码先定义一个类,然后使用 HttpClient 发送 POST 请求,传递要请求的数据,如 phone、paswd,然后取得返回的结果集。

**2. PHP 服务器端请求处理**

接下来需要写服务器端的程序,新建一个 user_login.php 文件,该文件首先连接数据库,然后接收用户传入的数据,最后根据用户传入的 phone 和 paswd 去匹配数据库中的数据来验证是否存在该用户。

由于需要操作的请求较多,如果每个请求都重写数据库连接,会造成代码冗余严重,所以把数据库连接单独放在 conn.php 文件下,需要时引用就可以了。具体操作如下。

数据库连接文件 conn.php 文件代码如下所示:

```php
$ servername = "localhost:3308";              //WampServer 中 MySQL 的端口号
$ username = "root";
$ password = "";
$ dbname = "user";
//创建连接
$ conn = mysqli_connect( $ servername, $ username, $ password, $ dbname);
//连接失败,打印错误信息
if (! $ conn) {
    echo 'error';
}
```

user_login.php 文件代码如下所示:

```php
<?php
    //登录验证
```

```php
include('conn.php');
//获取用户名和密码
$phone = $_POST["phone"];
$paswd = $_POST["paswd"];
$sql = "SELECT * FROM user_mes where phonenum = '$phone' and paswd = '$paswd'";
$result = $conn->query($sql);
if ($result->num_rows > 0) {
    echo 'success';
} else {
    echo "0";
}
$conn->close();
?>
```

上述代码中,第一句加粗的代码是代表与数据库建立连接,第二和第三句加粗代码表示接收客户端发送的数据,使用查询的方式来匹配是否存在该用户来决定是否允许用户登录。

### 3. 客户端调用请求发送类

客户端调用请求发送类时,传入要发送的请求参数和需要访问的URL。需要注意的是,在调用请求发送类时,要开启新线程或者使用第4章所讲的异步任务来操作,因为Android主线程是不允许进行耗时性的操作的,具体代码如下所示:

```java
//启动一个新的线程用来登录进行耗时操作
Runnable runnable = new Runnable() {
    @Override
    public void run() {
        HttpLogin httpLogin = new HttpLogin();
        String phone = phone_edit.getText().toString();
        String paswd = paswd_edit.getText().toString();
        //连接到服务器的地址,Android访问本机的IP是10.0.2.2
        String connectURL = 
        "http://192.168.56.1/teacher_pro/par_login.php";
        flag = http_Conn.gotoConn(phone, paswd, connectURL);
        if (flag) {
            Intent intent2 = new Intent(LoginActivity.this,
                com.example.teacher_pro.index.SucesActivity.class);
            Bundle bundle = new Bundle();
            bundle.putString("phone", phone);
            intent2.putExtras(bundle);
            startActivity(intent2);
        }else {
            Looper.prepare();
            Toast.makeText(LoginActivity.this, "登录失败,请重新登录", Toast.LENGTH_SHORT).show();
            Looper.loop();
        }
        
    }
};
```

以上就是客户端采用重新开启一个新线程的方式来发送网络请求,从而避免了主线程出现异常等问题。在上述代码中,第一句加粗的代码指定了要请求的 URL,第二句调用了发送请求类的请求方法。

以上就是使用 Android+PHP+MySQL 来实现 Android 前端与 PHP 服务器端交互。一般在处理用户的请求时,按照这样的步骤即可实现。

## 10.3.2 用户登录案例

在 10.3.1 节中,介绍了 Android+PHP+MySQL 结合使用的步骤,接下来通过一个用户登录的例子来更详细地讲解如何使用 PHP 作为后台服务器进行有关网络请求方面的应用程序开发。首先创建 Chapter10_user_login 项目,下面介绍具体实现。

### 1. 修改页面布局

创建完项目以后,修改界面布局代码。整体采用线性垂直布局的方式,然后放置两个 EditText 组件和一个 Button 组件,具体代码如下所示:

```xml
<?xml version = "1.0" encoding = "utf-8"?>
<LinearLayout xmlns:android = "http://schemas.android.com/apk/res/android"
    xmlns:tools = "http://schemas.android.com/tools"
    android:layout_width = "match_parent"
    android:layout_height = "match_parent"
    tools:context = "com.jxust.cn.chapter10_user_login.MainActivity"
    android:orientation = "vertical">
    <EditText
        android:id = "@ + id/phone"
        android:layout_width = "match_parent"
        android:layout_height = "wrap_content"
        android:hint = "请输入手机号"/>
    <EditText
        android:id = "@ + id/paswd"
        android:layout_width = "match_parent"
        android:layout_height = "wrap_content"
        android:inputType = "numberPassword"
        android:hint = "请输入密码"/>
    <Button
        android:id = "@ + id/login"
        android:layout_width = "match_parent"
        android:layout_height = "wrap_content"
        android:text = "登录"/>
</LinearLayout>
```

### 2. 创建客户端发送网络请求类

客户端请求类负责使用 HttpClient 发送网络请求,然后供按钮单击时调用。Http_Conn.java 的具体代码如下所示:

```
public class Http_Conn {
    //连接的方法
    public boolean gotoConn(String phonenum, String password, String connectUrl) {
        String result = "";                              //用来取得返回的String
        boolean isLoginSucceed = false;
        HttpClient httpClient = new DefaultHttpClient();
        //发送POST请求
        HttpPost httpRequest = new HttpPost(connectUrl);
        //POST运作传送变数必须用NameValuePair[]阵列存储
        List<NameValuePair> params = new ArrayList<NameValuePair>();
        //BasicNameValuePair存储键值对的类
        params.add(new BasicNameValuePair("phone", phonenum));
        params.add(new BasicNameValuePair("paswd", password));
        try {
            //发出HTTP请求转为带参数的HTTP网络地址
            httpRequest.setEntity(new UrlEncodedFormEntity(params,"utf-8"));
            //取得HTTP Response
            HttpResponse httpResponse = httpClient.execute(httpRequest);
            result = EntityUtils.toString(httpResponse.getEntity());

        } catch (Exception e) {
            e.printStackTrace();
        }
        //判断返回的数据是否为PHP中成功登录时输出的success
        if (result.equals("success")) {
            isLoginSucceed = true;
        }
        return isLoginSucceed;
    }
}
```

需要注意的是，Android 6.0以后就不能直接使用HttpClient了，这里需要在build.gradle(Module)里的android下添加一句代码，具体操作如图10-19所示。

图10-19 修改配置

### 3. 数据库数据添加

这里涉及的数据较少，就使用了手动添加的方式，在10.2节创建的user数据库中新建一个数据表user_login，然后添加一条数据。添加完后的结果如图10-20所示。

图 10-20　添加数据

### 4. 创建后台服务器 PHP 文件

接下来就是编写 PHP 文件，因为涉及数据库，所以首先要连接 user 数据库，然后取得客户端传来的 phone 和 paswd。连接数据库的文件 conn.php 代码如下所示：

```php
<?php
    $servername = "localhost:3308";
    $username = "root";
    $password = "";
    $dbname = "user";
    //创建连接
    $conn = mysqli_connect($servername, $username, $password, $dbname);
    //连接失败,打印错误信息
    if (!$conn) {
        echo 'error';
    }
?>
```

处理请求以及操作数据库的 user_login.php 文件代码如下所示：

```php
<?php
    include('conn.php');
    $phone = $_POST["phone"];
    $paswd = $_POST["paswd"];
    $sql = "SELECT * FROM user_login where phonenum = '$phone' and paswd = '$paswd'";
    $result = $conn->query($sql);
    if ($result->num_rows > 0) {
        echo 'success';
    } else {echo "0";
}
    $conn->close();
?>
```

### 5. 编写 MainActivity 代码

上述 PHP 文件编写完后，接着就需要编写 MainActivity 类的代码，该类负责初始化组件以及为按钮添加单击事件，发送网络请求，具体代码如下所示：

```java
public class MainActivity extends AppCompatActivity {
    private EditText phone_edit,paswd_edit;
    private Button login_btn;
    public Boolean flag = false;
    @Override
    protected void onCreate(Bundle savedInstanceState) {
        super.onCreate(savedInstanceState);
        setContentView(R.layout.activity_main);
        init();
    }
    //组件初始化方法
    private void init(){
        phone_edit = (EditText)findViewById(R.id.phone);
        paswd_edit = (EditText)findViewById(R.id.paswd);
        login_btn = (Button)findViewById(R.id.login);
        login_btn.setOnClickListener(new View.OnClickListener() {
            @Override
            public void onClick(View view) {
                //启动线程
                Thread thread = new Thread(runnable);
                thread.start();
            }
        });
    }
    //启动一个新的线程用来登录进行耗时操作
    Runnable runnable = new Runnable() {
        @Override
        public void run() {
            Http_Conn httpconn = new Http_Conn();
            String phone = phone_edit.getText().toString();
            String paswd = paswd_edit.getText().toString();
            //连接到服务器的地址,Android 访问本机的 IP 是 10.0.2.2
            String connectURL =
                    "http://10.0.2.2/user_login.php";
            flag = httpconn.gotoConn(phone, paswd, connectURL);
            if (flag) {
                //传入手机号用来在 me_layout 界面显示
                //成功后显示消息
                Looper.prepare();
                Toast.makeText(MainActivity.this,"登录成功",
                Toast.LENGTH_SHORT).show();
                Looper.loop();
            }else {
                Looper.prepare();
                Toast.makeText(MainActivity.this,"登录失败",
                Toast.LENGTH_SHORT).show();
                Looper.loop();
            }
```

```
            }
        };
    }
```

#### 6. 添加网络访问权限

由于需要与服务器进行交互,所以需要添加网络权限。清单中添加网络的代码如下所示:

```
<uses-permission android:name="android.permission.INTERNET"/>
<application
    ...
    android:usesCleartextTraffic="true"
>
<uses-library android:name="org.apache.http.legacy" android:required="false"/>
```

#### 7. 运行

运行程序,输入数据库中已经存在的数据,结果如图 10-21 所示。

图 10-21　运行结果

## 10.4　本章小结

本章主要讲解 PHP+MySQL+Android 的使用过程。首先讲解搭建 PHP 的开发环境的方法,PHP 作为服务器开发的优势。接着讲解如何将 PHP 与 MySQL 连接,以及连接 MySQL 以后,对数据库的各种操作等。最后通过一个用户登录的案例讲解客户端如何发送请求,以及服务器收到请求以后如何操作。本章的知识需要熟练掌握,在涉及数据库较多的开发场合,可以考虑使用 PHP 作为后台来操作数据库,把请求与视图隔离开来,简化 Android 应用程序。

## 10.5 课后习题

1. 使用PHP创建一个数据库,其中包含两个数据表。
2. 使用PHP往上述创建的两个数据表中分别添加一条数据。
3. 使用PHP+MySQL+Android实现用户的注册。

# 第11章

# "倾心家教"应用案例开发

近年来,由于传统的培训机构成本飞涨,使得传统线下教育面临巨大压力。而且传统培训机构模式存在着大量的问题,比如时间、地点上的限制,师资力量问题,不能满足个性化的需求,等等。与之相反的是,在线查找教师和预约教师有很强的可选择性,可以满足不同用户的不同需要,随时随地,想学就学,极大地增加了用户的可选择度,降低了教育所带来的成本。本章将对"倾心家教"应用开发的过程进行详细的讲解。

## 11.1 应用分析

本应用要实现的功能包括教师的查询、优秀教师在线显示、App所提供的课程、教师查询预约、建议反馈以及个人信息管理等。为了便于理解,接下来将通过一个用例图来说明,如图11-1所示。

图 11-1 应用程序用例图

从图 11-1 所示的用例图可以清楚地知道用户需要哪些功能,接下来针对这些功能对应的 UI 界面设计进行详细的讲解。

## 11.2 应用界面设计

### 11.2.1 登录界面设计

登录界面包括两个 EditText 组件、一个 CheckBox 组件、两个 Button 组件,分别用来输入手机号、密码、选择是否记住密码、登录或者注册,具体的界面设计如图 11-2 和图 11-3 所示。

图 11-2　登录界面　　　　　　图 11-3　注册界面

注册时,首先输入手机号、密码,然后单击"获取验证码"按钮,获取验证码并填写正确以后,才能进行注册,所以包含三个 EditText 组件、两个 Button 组件。

由于注册和登录的布局代码类似,这里只展示登录的布局代码。登录界面 login.layout 的具体代码如下所示:

```xml
<?xml version = "1.0" encoding = "utf-8"?>
< LinearLayout xmlns:android = "http://schemas.android.com/apk/res/android"
    android:layout_width = "match_parent"
    android:layout_height = "match_parent"
    android:background = " # EEEEEE"
    android:orientation = "vertical" >
  < TextView
      android:layout_width = "wrap_content"
      android:layout_height = "wrap_content"
      android:layout_gravity = "center_horizontal"
      android:layout_marginTop = "130dp"
      android:text = "LOGIN"
      android:textColor = " # 35A9D0"
```

```xml
            android:textSize = "25sp" />
<LinearLayout
    android:layout_width = "wrap_content"
    android:layout_height = "wrap_content"
    android:layout_gravity = "center_horizontal"
    android:layout_marginTop = "20dp"
    android:orientation = "horizontal" >
    <ImageView
        android:layout_width = "40dp"
        android:layout_height = "40dp"
        android:background = "#35A9D0"
        android:src = "@drawable/zhanghao" />
    <EditText
        android:id = "@ + id/login_phonenumber"
        android:layout_width = "200dp"
        android:layout_height = "40dp"
        android:background = "@drawable/logincontentshape"
        android:hint = "请输入手机号"
        android:maxLength = "11"
        android:maxLines = "1" />
</LinearLayout >
<LinearLayout
    android:layout_width = "wrap_content"
    android:layout_height = "wrap_content"
    android:layout_gravity = "center_horizontal"
    android:layout_marginTop = "15dp"
    android:orientation = "horizontal" >
    <ImageView
        android:layout_width = "40dp"
        android:layout_height = "40dp"
        android:background = "#35A9D0"
        android:src = "@drawable/mima" />
    <EditText
        android:id = "@ + id/login_paswd"
        android:layout_width = "200dp"
        android:layout_height = "40dp"
        android:background = "@drawable/logincontentshape"
        android:hint = "请输入密码"
        android:inputType = "textPassword"
        android:maxLength = "16"
        android:maxLines = "1" />
</LinearLayout >
<LinearLayout
    android:layout_width = "fill_parent"
    android:layout_height = "wrap_content"
    android:layout_marginTop = "10dp"
    android:orientation = "horizontal" >
    <CheckBox
```

```xml
            android:id = "@ + id/reme_paswd"
            android:layout_width = "0dp"
            android:layout_weight = "1"
            android:layout_height = "wrap_content"
            android:layout_marginLeft = "70dp"
            android:checked = "true"
            android:text = "记住密码"
            android:textColor = "#8B8B7A"
            android:textSize = "16sp" />
        <TextView
            android:id = "@ + id/forget_paswd"
            android:layout_width = "0dp"
            android:layout_weight = "1"
            android:layout_height = "wrap_content"
            android:layout_marginLeft = "20dp"
            android:text = "忘记密码"
            android:textColor = "#8B8B7A"
            android:textSize = "16sp" />
    </LinearLayout>
    <Button
        android:id = "@ + id/login"
        android:layout_width = "240dp"
        android:layout_height = "45dp"
        android:layout_gravity = "center_horizontal"
        android:layout_marginTop = "30dp"
        android:background = "#35A9D0"
        android:text = "登 录"
        android:textColor = "#FFFFFF"
        android:textSize = "22sp" />
    <LinearLayout
        android:layout_width = "fill_parent"
        android:layout_height = "wrap_content"
        android:layout_marginTop = "20dp"
        android:orientation = "vertical">
        <TextView
            android:id = "@ + id/regester_user"
            android:layout_width = "wrap_content"
            android:layout_height = "wrap_content"
            android:layout_gravity = "center_horizontal"
            android:text = "注 册"
            android:textColor = "#35A9D0"
            android:textSize = "18sp" />
    </LinearLayout>
</LinearLayout>
```

由于对应用的 UI 界面进行了美化,所以代码较长,其中的 logincontentshape 文件就是自定义的样式文件,开发者可以根据实际情况进行修改。

## 11.2.2 主界面规划设计

主界面默认显示的是底部导航栏首页所对应的界面，包含搜索教师框、课程图标和名称、滚动消息、优秀教师图片滚动以及底部导航栏，具体的设计如图 11-4 所示。

图 11-4 应用程序主界面设计

单击图 11-4 中的课程图标，即可显示当前这门课程的教师，底部的导航栏分别对应首页界面（程序主界面）、查询/预约教师界面、消息界面以及个人信息界面。

主界面包含的内容较多，首先顶部包含一个按钮和一个搜索框组件，接着是一块采用线性布局方式的课程图标和名字，然后就是滚动的消息和优秀的教师动态展示。其中滚动的消息和动态展示的教师图片是通过代码来实现的，这里就先讲解其他静态的布局方式。index.layout 的代码如下所示：

```xml
<?xml version = "1.0" encoding = "utf-8"?>
<LinearLayout
    xmlns:android = "http://schemas.android.com/apk/res/android"
    android:layout_width = "match_parent"
    android:layout_height = "match_parent"
    android:orientation = "vertical">
    <!-- 主要内容 -->
    <LinearLayout
        android:layout_width = "match_parent"
        android:layout_height = "0dp"
        android:layout_weight = "3"
        android:orientation = "vertical">
        <LinearLayout
            android:layout_width = "match_parent"
            android:layout_height = "52dp"
```

```xml
        android:background = "#11cd6e"
        android:orientation = "horizontal">
        <!-- 显示地点 -->
        <TextView
            android:id = "@+id/dizhi"
            android:layout_width = "0dp"
            android:layout_height = "30dp"
            android:layout_weight = "1"
            android:layout_marginTop = "14dp"
            android:layout_marginLeft = "10dp"
            android:textColor = "#FFFFFF"
            android:textSize = "18sp"
            android:text = "赣    州" />
        <!-- 显示搜索框 -->
        <SearchView
            android:id = "@+id/word"
            android:layout_width = "0dp"
            android:layout_height = "40dp"
            android:layout_weight = "4"
            android:layout_marginTop = "6dp"
            android:layout_marginRight = "10dp"
            android:background = "@drawable/search_shape"
            android:singleLine = "true" >
        </SearchView>
    </LinearLayout>
    <!-- 课程 -->
    <LinearLayout
        android:layout_width = "match_parent"
        android:layout_height = "155dp"
        android:background = "#FFFFFF"
        android:orientation = "vertical">
        <LinearLayout
            android:layout_width = "match_parent"
            android:layout_height = "70dp"
            android:layout_marginTop = "15dp"
            android:orientation = "horizontal">
            <!-- 显示课程图标和名称 -->
            <LinearLayout
            android:layout_width = "wrap_content"
            android:layout_height = "wrap_content"
            android:layout_weight = "1"
            android:layout_gravity = "center_horizontal"
            android:orientation = "vertical">
        <ImageView
          android:id = "@+id/math"
          android:layout_width = "50dp"
          android:layout_height = "50dp"
          android:layout_gravity = "center_horizontal"
```

```xml
        android:src = "@drawable/math"/>
<TextView
    android:layout_width = "wrap_content"
    android:layout_height = "wrap_content"
    android:layout_gravity = "center_horizontal"
    android:textSize = "14sp"
    android:textColor = "#8B8B7A"
    android:text = "数学" />
</LinearLayout>
    ...
</LinearLayout>
</LinearLayout>
<include layout = "@layout/index_toutiao_scroll" />
<include layout = "@layout/index_teacher" />
</LinearLayout>
    <!-- 热门名师 -->
        <include layout = "@layout/famous_teacher" />
</LinearLayout>
```

滚动的消息(index_toutiao_scroll)、头条教师文字(index_teacher)以及动态的优秀教师图片(famous_teacher)的布局采用了<include>的方式加载到主界面中。

index_toutiao_scroll 的代码如下所示：

```xml
<LinearLayout xmlns:android = "http://schemas.android.com/apk/res/android"
    xmlns:tools = "http://schemas.android.com/tools"
    android:id = "@+id/activity_main"
    android:layout_width = "match_parent"
    android:layout_height = "50dp"
    android:orientation = "vertical">
    <ViewFlipper
        android:id = "@+id/marquee_view"
        android:layout_width = "match_parent"
        android:layout_marginTop = "10dp"
        android:layout_height = "35dp"
        android:autoStart = "true"
        android:flipInterval = "2500"
        android:inAnimation = "@drawable/anim_marquee_in"
        android:outAnimation = "@drawable/anim_marquee_out" >
    </ViewFlipper>
</LinearLayout>
```

由于使用了 ViewFlipper 组件，因此通过动画的形式，实现布局文件的切换。其中 anim_marquee_in 的代码如下所示：

```xml
<?xml version = "1.0" encoding = "utf-8"?>
<set xmlns:android = "http://schemas.android.com/apk/res/android">
<translate
```

```xml
    android:duration = "1500"
    android:fromYDelta = "100 % p"
    android:toYDelta = "0">
</translate>
</set>
```

anim_marquee_out 的代码如下所示：

```xml
<?xml version = "1.0" encoding = "utf-8"?>
<set xmlns:android = "http://schemas.android.com/apk/res/android">
<translate
    android:duration = "1500"
    android:fromYDelta = "0"
    android:toYDelta = " - 100 % p">
</translate>
</set>
```

index_teacher 的代码如下所示：

```xml
<?xml version = "1.0" encoding = "utf-8"?>
<LinearLayout
    xmlns:android = "http://schemas.android.com/apk/res/android"
    android:orientation = "vertical"
    android:layout_width = "match_parent"
    android:background = "#FFFFFF"
    android:layout_marginTop = "6dp"
    android:layout_height = "40dp">
    <TextView
        android:layout_width = "wrap_content"
        android:layout_height = "35dp"
        android:layout_marginTop = "3dp"
        android:layout_gravity = "center_horizontal"
        android:textSize = "22sp"
        android:text = "热门名师 "/>
</LinearLayout>
```

famous_teacher 布局文件的代码如下所示：

```xml
<?xml version = "1.0" encoding = "utf-8"?>
<RelativeLayout xmlns:android = "http://schemas.android.com/apk/res/android"
    xmlns:tools = "http://schemas.android.com/tools"
    android:layout_width = "match_parent"
    android:layout_marginTop = "5dp"
    android:layout_height = "wrap_content" >
    <FrameLayout
        android:layout_width = "fill_parent"
        android:layout_height = "200dip"
```

```xml
<android.support.v4.view.ViewPager
    android:id = "@+id/vp"
    android:layout_width = "fill_parent"
    android:layout_height = "fill_parent" />
<LinearLayout
    android:layout_width = "fill_parent"
    android:layout_height = "35dip"
    android:orientation = "vertical"
    android:layout_gravity = "bottom"
    android:gravity = "center"
    android:background = "#33000000" >
    <TextView
        android:id = "@+id/image_title"
        android:layout_width = "wrap_content"
        android:layout_height = "wrap_content"
        android:text = "图片标题"
        android:textColor = "@android:color/white"
        />
    <LinearLayout
        android:layout_width = "wrap_content"
        android:layout_height = "wrap_content"
        android:orientation = "horizontal"
        android:layout_marginTop = "3dip" >
        <View
            android:id = "@+id/dot_0"
            android:layout_width = "5dip"
            android:layout_height = "5dip"
            android:layout_marginLeft = "2dip"
            android:layout_marginRight = "2dip"
            android:background = "@drawable/dot_focused"/>
        <View
            android:id = "@+id/dot_1"
            android:layout_width = "5dip"
            android:layout_height = "5dip"
            android:layout_marginLeft = "2dip"
            android:layout_marginRight = "2dip"
            android:background = "@drawable/dot_normal" />
        <View
            android:id = "@+id/dot_2"
            android:layout_width = "5dip"
            android:layout_height = "5dip"
            android:layout_marginLeft = "2dip"
            android:layout_marginRight = "2dip"
            android:background = "@drawable/dot_normal"/>
        <View
            android:id = "@+id/dot_3"
            android:layout_width = "5dip"
            android:layout_height = "5dip"
```

```xml
                        android:layout_marginLeft = "2dip"
                        android:layout_marginRight = "2dip"
                        android:background = "@drawable/dot_normal"/>
                    <View
                        android:id = "@ + id/dot_4"
                        android:layout_width = "5dip"
                        android:layout_height = "5dip"
                        android:layout_marginLeft = "2dip"
                        android:layout_marginRight = "2dip"
                        android:background = "@drawable/dot_normal"/>
                </LinearLayout>
        </LinearLayout>
    </FrameLayout>
</RelativeLayout>
```

以上就是主界面的布局文件代码。接下来讲解查找教师界面、信息界面以及个人信息界面的布局。

### 11.2.3 找教师界面

在找教师界面可根据用户选择的年级和科目的组合条件来查询,单击科目或者年级会弹出一个下拉菜单,用户可以选择要查找的年级或者科目。涉及的3个文件分别为 seach.xml、gradle_pouwin.xml 和 sub_pouwin.xml。

search.xml 是找教师的主界面,用户可以选择年级或科目,然后单击"查询"按钮。具体代码如下所示:

```xml
<?xml version = "1.0" encoding = "utf-8"?>
<LinearLayout
    xmlns:android = "http://schemas.android.com/apk/res/android"
    android:layout_width = "match_parent"
    android:layout_height = "match_parent"
    android:background = "#E8E8E8"
    android:orientation = "vertical">
    <!-- 显示要查找老师的信息标题 -->
    <LinearLayout
        android:layout_width = "match_parent"
        android:layout_height = "51dp"
        android:background = "#11cd6e"
        android:orientation = "horizontal">
    <!-- 返回箭头 -->
    <ImageView
        android:id = "@ + id/arrback"
        android:layout_width = "0dp"
        android:layout_weight = "0.5"
        android:layout_height = "30dp"
        android:layout_marginTop = "10dp"
```

```xml
        android:contentDescription = "@string/app_name"
        android:src = "@drawable/arrowleft" />
<!-- 科目文字 -->
<LinearLayout
    android:layout_width = "0dp"
    android:layout_weight = "5"
    android:layout_height = "50dp"
    android:orientation = "vertical">
    <TextView
        android:layout_width = "wrap_content"
        android:layout_gravity = "center"
        android:layout_height = "wrap_content"
        android:textSize = "22sp"
        android:textColor = "#FFFFFF"
        android:layout_marginTop = "12dp"
        android:text = "查找教师"/>
</LinearLayout>
</LinearLayout>
<!-- 间隔线 -->
<ImageView
    android:layout_width = "match_parent"
    android:layout_height = "1dp"
    android:background = "#CFCFCF" />
<LinearLayout
    android:layout_width = "match_parent"
    android:layout_height = "wrap_content"
    android:background = "#FFFFFF"
    android:orientation = "horizontal">
    <LinearLayout
        android:layout_width = "0dp"
        android:layout_weight = "1.5"
        android:layout_height = "wrap_content"
        android:orientation = "vertical">
        <Button
            android:id = "@+id/subject_name"
            android:layout_width = "wrap_content"
            android:layout_gravity = "center_horizontal"
            android:text = "科    目"
            android:textColor = "#8B8B7A"
            android:background = "@android:color/transparent"
            android:drawableRight = "@drawable/arrowdown"
            android:layout_height = "40dp"/>
    </LinearLayout>
<!-- 间隔线 -->
<ImageView
    android:layout_width = "1dp"
    android:layout_height = "30dp"
    android:layout_marginTop = "5dp"
```

```xml
            android:background = "#CFCFCF" />
        <LinearLayout
            android:layout_width = "0dp"
            android:layout_weight = "1.5"
            android:layout_height = "wrap_content"
            android:orientation = "vertical">
            <Button
                android:id = "@+id/grade"
                android:layout_width = "wrap_content"
                android:layout_gravity = "center_horizontal"
                android:text = "年级"
                android:textColor = "#8B8B7A"
                android:background = "@android:color/transparent"
                android:drawableRight = "@drawable/arrowdown"
                android:layout_height = "40dp"/>
        </LinearLayout>
    </LinearLayout>
    <!-- 间隔线 -->
    <ImageView
        android:layout_width = "match_parent"
        android:layout_height = "1dp"
        android:background = "#CFCFCF" />
    <ListView
        android:id = "@+id/select_teacher"
        android:layout_width = "fill_parent"
        android:layout_height = "wrap_content"
        android:background = "#FFFFFF">
    </ListView>
</LinearLayout>
```

上述代码中首先定义了科目和年级的 Button 组件,用户单击科目按钮时,会出现 gradle_pouwin.xml 所对应的年级选择界面。具体代码如下所示:

```xml
<?xml version = "1.0" encoding = "utf-8"?>
<LinearLayout
    xmlns:android = "http://schemas.android.com/apk/res/android"
    android:orientation = "vertical"
    android:layout_width = "fill_parent"
    android:background = "#ffffff"
    android:layout_marginTop = "1dp"
    android:layout_height = "380dp">
    <!-- 间隔线 -->
    <ImageView
        android:layout_width = "match_parent"
        android:layout_height = "1dp"
        android:background = "#CFCFCF" />
    <!-- 小学 -->
```

```xml
<LinearLayout
    android:orientation = "horizontal"
    android:layout_width = "fill_parent"
    android:layout_marginTop = "5dp"
    android:layout_height = "wrap_content">
    <!-- 一年级 -->
    <LinearLayout
        android:orientation = "vertical"
        android:layout_width = "0dp"
        android:layout_weight = "1"
        android:layout_height = "wrap_content">
        <Button
            android:id = "@+id/grade1"
            android:layout_width = "100dp"
            android:layout_height = "36dp"
            android:text = "一年级"
            android:layout_gravity = "center_horizontal"
            android:background = "@drawable/grade_edit1"
            android:textColor = "#8B8B7A"/>
    </LinearLayout>
        ...
</LinearLayout>
    <!-- 确定 -->
    <LinearLayout
        android:orientation = "vertical"
        android:layout_marginTop = "30dp"
        android:layout_width = "match_parent"
        android:layout_height = "wrap_content">
        <Button
            android:id = "@+id/sure_grade_select"
            android:layout_width = "fill_parent"
            android:layout_height = "40dp"
            android:textColor = "#FFFFFF"
            android:text = "确定"
            android:background = "#11cd6e"
            android:layout_gravity = "center"/>
    </LinearLayout>
</LinearLayout>
```

这里只列出了一部分代码,因为其他年级的布局代码与上述代码中定义的一年级的方式类似,所以不再详细列出。

当用户单击科目按钮时,会出现 sub_pouwin.xml 文件所对应的选择科目界面。具体代码如下所示:

```xml
<?xml version = "1.0" encoding = "utf-8"?>
<LinearLayout
    xmlns:android = "http://schemas.android.com/apk/res/android"
```

```xml
android:orientation = "vertical"
android:layout_width = "fill_parent"
android:background = "#ffffff"
android:layout_marginTop = "1dp"
android:layout_height = "380dp">
<!-- 间隔线 -->
< ImageView
    android:layout_width = "match_parent"
    android:layout_height = "1dp"
    android:background = "#CFCFCF" />
< LinearLayout
    android:orientation = "horizontal"
    android:layout_width = "fill_parent"
    android:layout_marginTop = "15dp"
    android:layout_height = "wrap_content">
    <!-- 语文 -->
    < LinearLayout
        android:orientation = "vertical"
        android:layout_width = "0dp"
        android:layout_weight = "1"
        android:layout_height = "wrap_content">
        < Button
            android:id = "@+id/sub1"
            android:layout_width = "100dp"
            android:layout_height = "36dp"
            android:text = "语文"
            android:layout_gravity = "center_horizontal"
            android:background = "@drawable/grade_edit1"
            android:textColor = "#8B8B7A"/>
    </LinearLayout >
        ...
    < LinearLayout
        android:orientation = "vertical"
        android:layout_width = "0dp"
        android:layout_weight = "1"
        android:layout_height = "wrap_content">
    </LinearLayout >
</LinearLayout >
    <!-- 确定 -->
        < LinearLayout
            android:orientation = "vertical"
            android:layout_marginTop = "30dp"
            android:layout_width = "match_parent"
            android:layout_height = "wrap_content">
            < Button
                android:id = "@+id/sure_sub_select"
                android:layout_width = "fill_parent"
                android:layout_height = "40dp"
```

```
                android:textColor = "#FFFFFF"
                android:text = "确定"
                android:background = "#11cd6e"
                android:layout_gravity = "center"/>
    </LinearLayout>
</LinearLayout>
```

以上就是选择科目或者年级时所弹出的界面对应的布局代码,当用户确定了选择的科目和年级以后,就可以把符合条件的教师查询出来,使用 ListView 组件以列表的方式显示。

### 11.2.4 消息界面

消息界面包含了"预约老师消息""我要提建议""消息论坛""优惠信息"查询等功能。本应用程序只实现了前面的教师消息查询、提建议的功能,所以只讲解消息主界面(mes.xml)、教师消息查询(mes_order_teach.xml)和提建议界面(mes_order_suggest.xml)的布局方式。消息界面如图 11-5 所示。

图 11-5 消息界面

图 11-5 所示界面对应的布局文件 mes.xml 的代码如下所示:

```
<?xml version = "1.0" encoding = "utf-8"?>
<LinearLayout
    xmlns:android = "http://schemas.android.com/apk/res/android"
    android:orientation = "vertical"
    android:layout_width = "match_parent"
    android:layout_height = "match_parent"
    android:background = "#E8E8E8">
    <!-- 消息标题 -->
    <LinearLayout
        android:layout_width = "match_parent"
        android:layout_height = "51dp"
        android:background = "#11cd6e"
        android:orientation = "horizontal">
        <!-- 返回箭头 -->
        <ImageView
            android:layout_width = "0dp"
            android:layout_weight = "0.5"
```

```xml
        android:layout_height = "30dp"
        android:layout_marginTop = "10dp"
        android:contentDescription = "@string/app_name"
        android:src = "@drawable/arrowleft" />
    <!-- 消息 -->
    <LinearLayout
        android:layout_width = "0dp"
        android:layout_weight = "5"
        android:layout_height = "50dp"
        android:orientation = "vertical">
        <TextView
            android:layout_width = "wrap_content"
            android:layout_gravity = "center"
            android:layout_height = "wrap_content"
            android:textSize = "22sp"
            android:textColor = "#FFFFFF"
            android:layout_marginTop = "12dp"
            android:text = "消息通知"/>
    </LinearLayout>
</LinearLayout>
<!-- 预约教师信息 -->
<LinearLayout
        android:layout_width = "fill_parent"
        android:layout_height = "45dp"
        android:background = "#FFFFFF"
        android:orientation = "horizontal" >
    <ImageView
        android:layout_width = "0dp"
        android:layout_weight = "0.5"
        android:layout_height = "wrap_content"
        android:layout_marginLeft = "10dp"
        android:layout_gravity = "center_vertical"
        android:contentDescription = "@string/app_name"
        android:src = "@drawable/teacher" />
    <TextView
        android:id = "@+id/order_teach"
        android:layout_width = "0dp"
        android:layout_weight = "5"
        android:layout_height = "wrap_content"
        android:layout_gravity = "center"
        android:layout_marginLeft = "15dp"
        android:text = "预约老师消息"
        android:textSize = "16sp" />
    <ImageView
        android:layout_width = "0dp"
        android:layout_weight = "1"
        android:layout_height = "wrap_content"
        android:layout_gravity = "center_vertical"
```

```
                android:contentDescription = "@string/app_name"
                android:src = "@drawable/arrowright" />
        </LinearLayout>
</LinearLayout>
```

由于"我要提建议""消息论坛""优惠信息"的布局方式与"预约老师信息"布局方式一样,因此不再列出。

mes_order_teach 显示已经预约的教师的信息。具体代码如下所示:

```
<?xml version = "1.0" encoding = "utf-8"?>
<LinearLayout
    xmlns:android = "http://schemas.android.com/apk/res/android"
    android:orientation = "vertical"
    android:layout_width = "match_parent"
    android:layout_height = "match_parent">
    <!-- 预约教师 -->
    <LinearLayout
        android:layout_width = "match_parent"
        android:layout_height = "51dp"
        android:background = "#11cd6e"
        android:orientation = "horizontal">
        <!-- 返回箭头 -->
        <ImageView
            android:layout_width = "0dp"
            android:layout_weight = "0.5"
            android:layout_height = "30dp"
            android:layout_marginTop = "10dp"
            android:contentDescription = "@string/app_name"
            android:src = "@drawable/arrowleft" />
        <!-- 消息 -->
        <LinearLayout
            android:layout_width = "0dp"
            android:layout_weight = "5"
            android:layout_height = "50dp"
            android:orientation = "vertical">
            <TextView
                android:layout_width = "wrap_content"
                android:layout_height = "wrap_content"
                android:textSize = "20sp"
                android:textColor = "#FFFFFF"
                android:layout_marginTop = "12dp"
                android:text = "我预约的教师"/>
        </LinearLayout>
    </LinearLayout>
    <ListView
        android:id = "@ + id/order_teach_list"
        android:layout_width = "match_parent"
```

```xml
        android:layout_height = "wrap_content"
        android:divider = "#CFCFCF"
        android:dividerHeight = "1dp"
        android:background = "#FFFFFF">
    </ListView>
</LinearLayout>
```

mes_order_suggest.xml 对应着消息界面的提建议界面,包含了用户输入要提建议的科目、内容和主题,其界面设计如图 11-6 所示。

图 11-6 提建议界面

其所对应的具体代码如下所示:

```xml
<?xml version = "1.0" encoding = "utf-8"?>
<LinearLayout>
    <!-- 消息标题 -->
    <LinearLayout
        android:layout_width = "match_parent"
        android:layout_height = "51dp"
        android:background = "#11cd6e"
        android:orientation = "horizontal">
        <!-- 返回箭头 -->
        <ImageView
            android:layout_width = "0dp"
            android:layout_weight = "0.5"
            android:layout_height = "30dp"
            android:layout_marginTop = "10dp"
            android:contentDescription = "@string/app_name"
            android:src = "@drawable/arrowleft" />
        <!-- 消息 -->
        <LinearLayout
            android:layout_width = "0dp"
            android:layout_weight = "5"
            android:layout_height = "50dp"
            android:orientation = "vertical">
```

```xml
        <TextView
            android:layout_width="wrap_content"
            android:layout_gravity="center"
            android:layout_height="wrap_content"
            android:textSize="22sp"
            android:textColor="#FFFFFF"
            android:layout_marginTop="12dp"
            android:text="我的建议"/>
    </LinearLayout>
</LinearLayout>
    <!-- 建议科目 -->
    <LinearLayout
        android:layout_width="fill_parent"
        android:layout_height="50dp"
        android:background="#FFFFFF"
        android:orientation="horizontal">
        <TextView
            android:layout_width="0dp"
            android:layout_weight="1"
            android:layout_height="40dp"
            android:textSize="17sp"
            android:layout_marginTop="5dp"
            android:layout_marginLeft="20dp"
            android:text="科目:"/>
        <Spinner
            android:id="@+id/sugest_subject"
            android:layout_marginTop="5dp"
            android:layout_width="0dp"
            android:layout_weight="4"
            android:background="@drawable/spinner_shape"
            android:layout_marginRight="10dp"
            android:layout_height="40dp"/>
        ...
        <Button
            android:id="@+id/sugest_report"
            android:layout_width="fill_parent"
            android:layout_height="40dp"
            android:textColor="#FFFFFF"
            android:text="发表建议"
            android:background="#11cd6e"
            android:layout_gravity="center"/>
</LinearLayout>
```

以上就是消息界面的主要布局代码,开发者可以选择把顶部的布局文件单独列出来,然后导入每个布局文件即可使用。

### 11.2.5 个人信息界面

个人信息界面包含了用户的头像、用户的手机号、订单信息、钱包信息、奖学券信息、我已经预约的老师、学习计划与总结、我的课程以及系统的设置等功能,其实现效果如图 11-7 所示。

图 11-7 个人信息界面

个人信息界面中已经实现的功能有我的订单、我的钱包、奖学券、学习计划与总结、设置等,其中我的订单、钱包、奖学券布局代码都相对简单,此处不再赘述。下面主要讲解主界面的布局。整体采用线性垂直的布局方式,然后每一个功能采用线性水平的布局方式。

me.xml 文件的具体布局代码如下所示:

```
<LinearLayout>
    <LinearLayout
        android:id = "@ + id/user_message"
        android:layout_width = "fill_parent"
        android:layout_height = "110dp"
        android:background = "#11cd6e"
        android:orientation = "horizontal" >
        < ImageView
            android:id = "@ + id/icon"
            android:layout_width = "0dp"
            android:layout_height = "70dp"
            android:layout_weight = "1"
            android:layout_gravity = "center"
            android:layout_marginLeft = "10dp"
            android:contentDescription = "@string/app_name"
            android:src = "@drawable/hugh" />
```

```xml
            <TextView
                android:id = "@+id/phone_numbers"
                android:layout_width = "0dp"
                android:layout_weight = "4"
                android:layout_height = "wrap_content"
                android:layout_gravity = "center"
                android:text = "手机号:"
                android:textColor = "#FFFFFF"
                android:layout_marginLeft = "30dp"
                android:textSize = "20sp" />
            <ImageView
                android:layout_width = "0dp"
                android:layout_weight = "1"
                android:layout_height = "40dp"
                android:layout_gravity = "center_vertical"
                android:contentDescription = "@string/app_name"
                android:src = "@drawable/toparrowright" />
    </LinearLayout>
    <include layout = "@layout/me_money" />
    <LinearLayout
        android:layout_width = "fill_parent"
        android:layout_height = "90dp"
        android:layout_marginTop = "20dp"
        android:background = "#FFFFFF"
        android:orientation = "vertical" >
        <ImageView
            android:layout_width = "fill_parent"
            android:layout_height = "0.5dip"
            android:layout_marginLeft = "15dp"
            android:layout_marginRight = "15dp"
            android:background = "#CFCFCF" />
    </LinearLayout>
</LinearLayout>
```

以上省略了一部分功能的代码，其中加粗的那一行代码为我的订单、钱包以及奖学券的布局文件。me_money.xml 文件的布局代码如下所示：

```xml
<?xml version = "1.0" encoding = "utf-8"?>
<LinearLayout
    xmlns:android = "http://schemas.android.com/apk/res/android"
    android:orientation = "horizontal"
    android:layout_width = "fill_parent"
    android:background = "#FFFFFF"
    android:layout_height = "70dp">
    <!-- 我的订单 -->
    <LinearLayout
        android:orientation = "vertical"
```

```
            android:layout_width = "0dp"
            android:layout_weight = "1"
            android:layout_height = "match_parent">
        < ImageView
                android:id = "@ + id/order"
                android:layout_width = "wrap_content"
                android:layout_height = "wrap_content"
                android:layout_gravity = "center"
                android:layout_marginTop = "5dp"
                android:contentDescription = "@string/app_name"
                android:src = "@drawable/order" />
        < TextView
                android:layout_width = "wrap_content"
                android:layout_height = "wrap_content"
                android:layout_gravity = "center"
                android:text = "我的订单"
                android:layout_marginTop = "2dp"
                android:textSize = "16sp" />
    </LinearLayout>
    ...
</LinearLayout>
```

以上就是我的订单、钱包以及奖学券的布局代码,由于钱包和奖学券的布局方式与我的订单的布局方式一样,因此不再详细地列出来。

视频讲解

## 11.3 数据库设计

在进行应用程序的功能实现之前,首先需要设计应用程序所使用到的数据库。根据系统的用例图,设计了如图 11-8 所示的数据库。

数据库说明:

teache_pro——数据库名称。

evaluate——用户反馈的意见表。

parent——存储家长用户表(本应用面对的是家教家长客户端)。

par_money——家长钱包信息表。

par_order——家长订单表。

par_reward——拥有的奖学券信息表。

plant——个人信息界面学习计划对应的表。

reserve——对应家长预约的教师信息表。

teacher——教师信息表。

接下来将详细地介绍每个数据表所包含的信息。

图 11-8 数据库的结构

**1. parent: 家长用户表**

家长用户表主要包含:id、手机号、密码、头像以及地址信息,具体的结构如图 11-9 所示。

图 11-9 家长用户表属性图

**2. teacher：教师信息表**

教师信息表中包含 id、教师手机号、教师性别、教师的名字、教师的头像、地址、教学经历以及所教年级，具体如图 11-10 所示。

图 11-10 教师信息表属性图

**3. par_money：家长钱包信息表**

家长钱包信息表主要包含 id、家长手机号、账户的钱、余额以及积分信息，具体如图 11-11 所示。

图 11-11 家长钱包信息表属性图

**4. par_order：家长订单表**

家长订单表包含的信息有 id、家长手机号、所购买图书的名称、数量、价格以及是否已经收货信息，具体属性如图 11-12 所示。

图 11-12　家长订单表属性图

### 5. par_reward：奖学券信息表

奖学券信息表里存储的是系统发给用户的奖学券信息，包含的属性有 id、家长手机号、奖学券面额以及数量信息，具体如图 11-13 所示。

图 11-13　奖学券信息表属性图

### 6. plant：学习计划表

学习计划表是用户用来存储计划学习的内容的表，包含的属性有 id、家长手机号、计划学习的内容以及计划完成的时间信息，具体如图 11-14 所示。

图 11-14　学习计划表属性图

### 7. reserve：预约教师信息表

预约教师信息表中存储的是家长预约教师的信息，包含的属性有 id、家长手机号、教师手机号、教师的名字、预约的科目以及预约的时间，具体如图 11-15 所示。

### 8. evaluate：反馈意见表

反馈意见表存储的是家长对平台所提供的科目任课老师或者其他提出的意见等信息，包含的属性有 id、家长手机号、教师手机号、科目、提出意见的内容以及时间信息，具体如图 11-16 所示。

图 11-15 预约教师信息表属性图

图 11-16 反馈意见表属性图

## 11.4 应用功能实现

在完成了界面设计和数据库设计以后,接下来就需要实现各个功能了,本应用程序所提供的功能有:登录/注册、主界面信息展示、教师查询/预约、个人信息的管理以及设计功能。由于在第10章已经讲解过如何登录,所以在这里就不再讲解登录功能是如何实现的。接下来讲详细地讲解其他功能的实现过程。

### 11.4.1 应用主界面实现

应用主界面是指用户登录或者注册成功以后进入的界面,界面的布局文件是 index.layout.xml,里面包含的功能有动态消息的展示、优秀教师图片的滚动显示。接下来分别介绍这两个功能如何实现的。

视频讲解

**1. 动态消息的展示**

动态消息的滚动显示使用的 ViewFlipper 组件,ViewFlipper 是一个切换组件,一般用于图片的切换,当然它是可以添加 View 的,而不限定只用于 ImageView,还可以自定义 View,只是经常用 ViewFlipper 来实现的是 ImageView 的切换,如果切换自定义的 View,倒不如使用 ViewPager 来做。

本应用程序使用静态的布局文件来显示滚动消息的内容。每一个布局文件(index_toutiao、index_toutiao1~index_toutiao3)的代码如下所示:

```xml
<?xml version = "1.0" encoding = "utf-8"?>
<LinearLayout
    xmlns:android = "http://schemas.android.com/apk/res/android"
    android:orientation = "horizontal"
    android:layout_width = "match_parent"
    android:layout_marginTop = "10dp"
    android:background = "#FFFFFF"
    android:layout_height = "35dp">
    <ImageView
        android:layout_width = "0dp"
        android:layout_height = "30dp"
        android:layout_weight = "2"
        android:src = "@drawable/qing"/>
    <TextView
        android:id = "@+id/scroll_mes"
        android:layout_width = "0dp"
        android:layout_height = "wrap_content"
        android:layout_weight = "7"
        android:layout_marginLeft = "15dp"
        android:textSize = "16sp"
        android:textColor = "#000000"
        android:layout_gravity = "center_vertical"
        android:text = "名师一对一辅导啦"/>
    <ImageView
        android:layout_width = "0dp"
        android:layout_height = "18dp"
        android:layout_weight = "1"
        android:layout_gravity = "center_vertical"
        android:src = "@drawable/arrowright"/>
</LinearLayout>
```

接着把这4个布局文件添加到ViewFlipper当中，由于在布局文件index_toutiao_scroll.xml文件中已经设置了自动开始的属性为true，所以添加完运行程序即可实现消息滚动显示，具体代码如下所示：

```
//消息滚动通知
vFlipper = (ViewFlipper)index.findViewById(R.id.marquee_view);
vFlipper.addView(View.inflate(getApplicationContext(), R.layout.index_toutiao, null));
vFlipper.addView(View.inflate(getApplicationContext(), R.layout.index_toutiao1, null));
vFlipper.addView(View.inflate(getApplicationContext(), R.layout.index_toutiao2, null));
vFlipper.addView(View.inflate(getApplicationContext(), R.layout.index_toutiao3, null));
```

以上就是消息滚动显示的实现过程，接下来介绍优秀教师图片滚动显示的实现原理。

### 2. 优秀教师图片的滚动显示

优秀教师图片的滚动显示的实现过程是：首先把图片按照存储时的id添加到一个整型数组中，然后定义内部类ViewPagerAdapter继承自PagerAdapter，然后重写PagerAdapter当中的方法，最后再定义一个类，开启新的线程，以及在 Activity 的生命周期方法 onCreate() 中设置切换图片的间隔时间和使用 Handler 来发送消息，具体实现代码如下：

```java
//热门名师
    //图片 id
    imageIds = new int[]{
            R.drawable.a,
            R.drawable.b,
            R.drawable.c,
            R.drawable.d,
            R.drawable.e
        };
    //图片标题
    titles = new String[]{
            "热门心理老师",
            "热门英语老师",
            "热门物理老师",
            "热门数学老师"
        };
    //显示的图片
    images = new ArrayList<ImageView>();
    for(int i = 0; i < imageIds.length; i++){
        ImageView imageView = new ImageView(this);
        imageView.setBackgroundResource(imageIds[i]);
        images.add(imageView);
    }
    //显示的点
    dots = new ArrayList<View>();
    dots.add(findViewById(R.id.dot_0));
    dots.add(findViewById(R.id.dot_1));
    dots.add(findViewById(R.id.dot_2));
    dots.add(findViewById(R.id.dot_3));
    dots.add(findViewById(R.id.dot_4));
    title = (TextView)index.findViewById(R.id.image_title);
    title.setText(titles[0]);
    mViewPager = (ViewPager)index.findViewById(R.id.vp);
    adapter = new ViewPagerAdapter();
    mViewPager.setAdapter(adapter);
    mViewPager.setOnPageChangeListener(new OnPageChangeListener() {
        @Override
        public void onPageSelected(int position) {
            // TODO Auto-generated method stub
            title.setText(titles[position]);
            oldPosition = position;
            currentItem = position;
        }
        @Override
        public void onPageScrolled(int arg0, float arg1, int arg2) {
```

```java
            // TODO Auto-generated method stub
        }
        @Override
        public void onPageScrollStateChanged(int arg0) {
            // TODO Auto-generated method stub
        }
    });
    private class ViewPagerAdapter extends PagerAdapter {
        public int getCount() {
            return images.size();
        }
        @Override
        public boolean isViewFromObject(View arg0, Object arg1) {
            // TODO Auto-generated method stub
            return arg0 == arg1;
        }
        @Override
        public void destroyItem(ViewGroup view, int position, Object object) {
            // TODO Auto-generated method stub
            view.removeView(images.get(position));
        }
        @Override
        public Object instantiateItem(ViewGroup view, int position) {
            // TODO Auto-generated method stu
            view.addView(images.get(position));
            return images.get(position);
        }
    }
    @Override
    protected void onStart() {
        super.onStart();
        scheduledExecutorService = Executors.newSingleThreadScheduledExecutor();
            //每隔2s切换一张图片
    scheduledExecutorService.scheduleWithFixedDelay(new ViewPagerTask(), 2, 4, TimeUnit.SECONDS);
    }
    //切换图片
    private class ViewPagerTask implements Runnable {
        @Override
        public void run() {
            // TODO Auto-generated method stub
            currentItem = (currentItem + 1) % imageIds.length;
            //更新界面
            handler.obtainMessage().sendToTarget();
        }
    }
    private Handler handler = new Handler(){
        @Override
        public void handleMessage(Message msg){
```

```
        // TODO Auto-generated method stub
        //设置当前页面
        mViewPager.setCurrentItem(currentItem);
        }
    };
    @Override
    protected void onStop(){
        super.onStop();
    }
```

以上就是滚动图片实现的全部代码,由于主界面需要包含4个主要的布局文件,分别为 index.layout、search.layout、mes.layout 和 me.layout,所以在主界面对应的 Activity 中需要先使用 ViewPagger 把4个布局文件加载进来,然后对各个布局文件中的组件进行初始化以及添加监听事件。在主界面对应的 Activity 中需要编写的代码较多,其对应的 SucesActivity 的部分代码如下所示:

```
public class SucesActivity extends Activity implements OnClickListener, OnTouchListener {
    private ViewPager ViewPager;
    private ImageButton index_image, search_image, mes_image, me_image;
    //index 里的组件
    public TextView local_text;
    public ImageView math, chinese, english, physical, politics, chemistry, biology, geography;
    private ViewFlipper vFlipper;
    //消息里的组件
    public TextView order_teach, order_sug, order_talk, order_cheap;
    PopupWindow cheap_window;
    //me 里的组件初始化
    public LinearLayout use_mes;
    public TextView phone, plant, set;
    public ImageView icon_img, order_img, money_img, reward_img;
    public String phonenum = "";
    /*
     * 用来设置左右来回滑动
     */
    private List<View> lists = new ArrayList<View>();
    private MyAdapter myAdapter;
    //下面4个按钮,用来改变背景颜色
    private LinearLayout layout1, layout2, layout3, layout4;
    //图片轮播组件
    public int imageIds[];
    public String[] titles;
    public ArrayList<ImageView> images;
    public ArrayList<View> dots;
    public TextView title;
    public ViewPager mViewPager;
    public ViewPagerAdapter adapter;
```

```java
public int oldPosition = 0;                    //记录上一次点的位置
public int currentItem;                         //当前页面
private ScheduledExecutorService scheduledExecutorService;
@Override
protected void onCreate(Bundle savedInstanceState) {
    super.onCreate(savedInstanceState);
    this.requestWindowFeature(Window.FEATURE_NO_TITLE);
    setContentView(R.layout.activity_index);
    Intent intent = this.getIntent();
    Bundle bundle = intent.getExtras();
    phonenum = bundle.getString("phone");
    init();
}
//组件初始化
@SuppressWarnings("deprecation")
public void init(){
    index_image = (ImageButton)findViewById(R.id.index);
    search_image = (ImageButton)findViewById(R.id.search);
    mes_image = (ImageButton)findViewById(R.id.xiaoxi);
    me_image = (ImageButton)findViewById(R.id.me);
    //加载4个Layout,用来设置背景
    layout1 = (LinearLayout)findViewById(R.id.layout1);
    layout2 = (LinearLayout)findViewById(R.id.layout2);
    layout3 = (LinearLayout)findViewById(R.id.layout3);
    layout4 = (LinearLayout)findViewById(R.id.layout4);
    //加载对应的布局文件
    View index = getLayoutInflater().inflate(R.layout.index, null);
    View search = getLayoutInflater().inflate(R.layout.search, null);
    View mes = getLayoutInflater().inflate(R.layout.mes, null);
    View me = getLayoutInflater().inflate(R.layout.me, null);
    lists.add(index);
    lists.add(search);
    lists.add(mes);
    lists.add(me);
    myAdapter = new MyAdapter(lists);
    ViewPager = (ViewPager)findViewById(R.id.viewPager);
    ViewPager.setAdapter(myAdapter);
    //设置底部按钮监听事件
    index_image.setOnClickListener(this);
    index_image.setOnTouchListener(this);
    search_image.setOnClickListener(this);
    search_image.setOnTouchListener(this);
    mes_image.setOnClickListener(this);
    mes_image.setOnTouchListener(this);
    me_image.setOnClickListener(this);
    me_image.setOnTouchListener(this);
    //index的组件
    local_text = (TextView)index.findViewById(R.id.dizhi);
```

```java
        local_text.setOnClickListener(this);
        //课程组件
        math = (ImageView)index.findViewById(R.id.math);
        math.setOnClickListener(this);
        //...(课程图标初始化代码,与上面一致,不再列出)
        //消息滚动通知(此处代码省略,前面已经讲过)
        //热门名师(此处代码省略,前面已经讲过)
        //search 的组件
        //message 的组件
        order_teach = (TextView)mes.findViewById(R.id.order_teach);
        order_sug = (TextView)mes.findViewById(R.id.order_suggest);
        order_talk = (TextView)mes.findViewById(R.id.order_talk);
        order_cheap = (TextView)mes.findViewById(R.id.order_cheaper);
        order_teach.setOnClickListener(this);
        order_sug.setOnClickListener(this);
        order_talk.setOnClickListener(this);
        order_cheap.setOnClickListener(this);
        //me 的组件
        use_mes = (LinearLayout)me.findViewById(R.id.user_message);
        use_mes.setOnClickListener(this);
        phone = (TextView)me.findViewById(R.id.phone_numbers);
        phone.setText("手机号:" + phonenum);
        order_img = (ImageView)me.findViewById(R.id.order);
        order_img.setOnClickListener(this);
        money_img = (ImageView)me.findViewById(R.id.money);
        money_img.setOnClickListener(this);
        reward_img = (ImageView)me.findViewById(R.id.reward);
        reward_img.setOnClickListener(this);
        plant = (TextView)me.findViewById(R.id.plant);
        plant.setOnClickListener(this);
        set = (TextView)me.findViewById(R.id.sets);
        set.setOnClickListener(this);
}
@Override
public void onClick(View v) {
    // TODO Auto-generated method stub
    switch (v.getId()) {
    //首页按钮
    case R.id.index:
        ViewPager.setCurrentItem(0);
        break;
    /* 首页组件监听事件,跳转到各科教师显示界面 */
    case R.id.math:
        Intent intent11 = new Intent(SucesActivity.this, Showtea_Activity.class);
        Bundle bundle11 = new Bundle();
        bundle11.putString("subject", "数学");
        bundle11.putString("phone", phonenum );
        intent11.putExtras(bundle11);
```

```java
            startActivity(intent11);
            break;
    //...此处代码与上边代码处理流程一样,不再详细列出
    //找教师
        case R.id.search:
            Intent intent = new Intent(SucesActivity.this,
                com.example.teacher_pro.search.SchTeActivity.class);
            startActivity(intent);
            break;
    //消息
        case R.id.xiaoxi:
            ViewPager.setCurrentItem(2);
            break;
    //消息界面的预约教师信息
    //消息界面的我的建议
            break;
    //消息界面的论坛(代码省略,Activity 跳转)
    //消息界面的优惠信息(代码省略,Activity 跳转)
    //我
        case R.id.me:
            ViewPager.setCurrentItem(3);
            break;
    //me_order(我的钱包、订单、奖学券处理 Activity)
    //me_money
    //me_reward
        default:
            break;
    }
}
@Override
public boolean onTouch(View v, MotionEvent event) {
    // TODO Auto-generated method stub
    switch (v.getId()) {
    //首页按钮
        case R.id.index:
            if (event.getAction() == MotionEvent.ACTION_DOWN) {
                layout1.setBackgroundColor(Color.rgb(152, 251, 152));
            } else if (event.getAction() == MotionEvent.ACTION_UP) {
                layout1.setBackgroundColor(Color.parseColor("#F5F5F5"));
            }
            break;
    //查找教师
        case R.id.search:
            if (event.getAction() == MotionEvent.ACTION_DOWN) {
                layout2.setBackgroundColor(Color.rgb(152, 251, 152));
            } else if (event.getAction() == MotionEvent.ACTION_UP) {
                layout2.setBackgroundColor(Color.parseColor("#F5F5F5"));
            }
```

```java
            break;
        //消息
        case R.id.xiaoxi:
            if (event.getAction() == MotionEvent.ACTION_DOWN) {
                layout3.setBackgroundColor(Color.rgb(152, 251, 152));
            } else if (event.getAction() == MotionEvent.ACTION_UP) {
                layout3.setBackgroundColor(Color.parseColor("#F5F5F5"));
            }
            break;
        //我
        case R.id.me:
            if (event.getAction() == MotionEvent.ACTION_DOWN) {
                layout4.setBackgroundColor(Color.rgb(152, 251, 152));
            } else if (event.getAction() == MotionEvent.ACTION_UP) {
                layout4.setBackgroundColor(Color.parseColor("#F5F5F5"));
            }
            break;
        default:
            break;
    }
    return false;
}
//优秀教师图片滚动显示(后面自定实现ViewPagger以后的代码,上面讲过)
private class ViewPagerAdapter extends PagerAdapter {
    public int getCount() {
        return images.size();
    }
    @Override
    public boolean isViewFromObject(View arg0, Object arg1) {
        // TODO Auto-generated method stub
        return arg0 == arg1;
    }
    @Override
    public void destroyItem(ViewGroup view, int position, Object object) {
        // TODO Auto-generated method stub
        view.removeView(images.get(position));
    }
    @Override
    public Object instantiateItem(ViewGroup view, int position) {
        // TODO Auto-generated method stu
        view.addView(images.get(position));
        return images.get(position);
    }
}
@Override
protected void onStart() {
    super.onStart();
}
```

```
//切换图片
private class ViewPagerTask implements Runnable {
    @Override
    public void run() {
        // TODO Auto-generated method stub
        currentItem = (currentItem + 1) % imageIds.length;
        //更新界面
        handler.obtainMessage().sendToTarget();
    }
}
private Handler handler = new Handler(){
    @Override
    public void handleMessage(Message msg){
        // TODO Auto-generated method stub
        //设置当前页面
        mViewPager.setCurrentItem(currentItem);
    }
};
```

从上述代码中可以看到,在进入主界面时,Activity 已经对导航栏中的 4 个功能所对应的界面进行了加载,并且对各个界面中的组件进行了初始化,然后为需要处理的组件添加了监听事件,例如课程图标的单击、底部导航栏的单击等。当用户单击课程图标或者底部的查询教师时,即可进入到查询教师所对应的 Activity。

以上就是主界面 Activity 所对应的部分代码,全部的代码可以在项目下的 SucesActivity 中查看。

### 11.4.2 教师查询/预约功能实现

在 11.4.1 节中讲到了单击首页的课程图标或者底部导航栏的"找教师"按钮时,会出现查询教师对应的界面,接下来就对查询教师以及教师预约功能进行详细的讲解。

#### 1. 教师查询

查询教师功能包括根据科目来查找和根据科目和年级的组合条件来查找两种方法。第一种是在面对用户单击首页上的课程图标时来处理的,这里主要讲的是第二种,即组合条件的查询。

查询教师的页面包含了 3 个布局文件,分别为 search.xml、sub_pouwin.xml,以及 grade_pouwin.xml,这 3 个文件的布局代码前面已经讲过,此处不再赘述。接着是处理用户选择年级和科目的组合条件事件,其对应的 Activity 为 SchTeActivity,首先对布局文件当中的组件进行初始化,然后为其添加监听事件,例如单击"找教师"按钮时,会出现选择科目或年级界面,用户可以根据科目和年级的组合条件来查找老师,然后预约。其对应的部分代码如下所示:

```
public class SchTeActivity extends Activity implements OnClickListener,OnItemClickListener {
    //获取科目和年级按钮
    //"返回"按钮
```

```java
//定义两个弹出框:课程和年级
//sub_pouwin 里的课程按钮
//grade_pouwin 里的年级按钮
//保存选择的年级和课程(默认的年级和课程)
private String sub_select = "语文", grade_select = "一年级", sub_select1 = "chinese";
//"确定"按钮
//显示教师列表 Listview
private ListView select_list_teach;
//暂时存储教师姓名
//存储教师信息列表
private String string_teach[] = new String[20];
private Boolean isget = false;
get_Select_Teach select = new get_Select_Teach();
@Override
protected void onCreate(Bundle savedInstanceState) {
    super.onCreate(savedInstanceState);
    this.requestWindowFeature(Window.FEATURE_NO_TITLE);
    setContentView(R.layout.search);
    init();
}
//组件初始化
public void init(){
}
@Override
public void onClick(View v) {
    switch (v.getId()) {
    case R.id.arrback:
    //返回首页
        break;
    case R.id.grade:
        //显示年级
        gradePoWin();
        break;
    case R.id.subject_name:
        //显示课程
        subPoWin();
        break;
    //选择的课程,其他的都类似于 sub1
    case R.id.sub1:
        sub_select = "语文";
        sub_select1 = "chinese";
        sub_btn.setText(sub_select);
        set_subBack();
        sub1.setBackgroundResource(R.drawable.grade_edit1);
        break;
    //确认选择的课程
    case R.id.sure_sub_select:
        subpoWind.dismiss();
```

```java
            break;
        //年级选择按钮(其他年级代码类似)
        case R.id.grade1:
            grade_select = "一年级";
            grade_btn.setText(grade_select);
            set_gradeBack();
            grade1.setBackgroundResource(R.drawable.grade_edit1);
            break;
        case R.id.sure_grade_select:
            //在这里加载数据库信息,显示出符合条件的教师信息
            gradepoWind.dismiss();
            new Anothertask().execute((Void[]) null);
            break;
        default:
            break;
    }
}
//显示年级 PopupWindow
protected void subPoWin() {
    //加载 PopupWindow 的布局文件
    contentView = LayoutInflater.from(getApplicationContext()).inflate(R.layout.sub_pouwin, null);
    //声明一个弹出框
    subpoWind = new PopupWindow(this.getWindowManager().getDefaultDisplay().getWidth(), 470);
    subpoWind.setContentView(contentView);
    //显示
    subpoWind.showAsDropDown(sub_btn);
    sub_init();
}
//显示年级 PopupWindow
protected void gradePoWin() {
    //加载 PopupWindow 的布局文件
    contentView1 = LayoutInflater.from(getApplicationContext()).inflate(
                                                R.layout.grade_pouwin, null);
    //声明一个弹出框
    gradepoWind = newPopupWindow(this.getWindowManager().getDefaultDisplay().getWidth(), 515);
    gradepoWind.setContentView(contentView1);
    //显示
    gradepoWind.showAsDropDown(grade_btn);
    grade_init();
}
//sub——pouwin 里的组件初始化
protected void sub_init() {
}
//grade——pouwin 里的组件初始化
protected void grade_init() {
    }
    //设置课程选择的背景颜色
    protected void set_subBack() {
```

```java
        sub1.setBackgroundResource(R.drawable.grade_edit);
        ...
    }
    //设置班级的选择颜色
    protected void set_gradeBack() {
        grade1.setBackgroundResource(R.drawable.grade_edit);
        ...
    }
    //异步任务获取教师信息
    private class Anothertask extends AsyncTask<Void, Integer, Boolean>{
        @Override
        protected Boolean doInBackground(Void... params) {
            //对UI组件的更新操作,是耗时的操作
            try {
                //连接到服务器的地址
        String connectURL = "http://192.168.56.1/teacher_pro/get_select_teach.php";
                //输入用户名密码和连接地址
                isget = select.get_Teach(sub_select1, grade_select, connectURL);
            } catch (Exception e) {
                e.printStackTrace();
            }
            return null;
        }
        @Override
        protected void onPostExecute(Boolean result) {
            //TODO Auto-generated method stub
            if (isget) {
                string_teach = select.result.split(",");
                select_list_teach.setAdapter(new ArrayAdapter<String>
                  (SchTeActivity.this,R.layout.array_adapt,string_teach));
            }else {
Toast.makeText(SchTeActivity.this, "抱歉,没有满足你要查找的教师!", Toast.LENGTH_SHORT).show();
            }
        }
    }
    @Override
    public void onItemClick(AdapterView<?> parent, View view, int position, long id) {
        //传递教师姓名,用来获取对应教师具体信息
        teach_name = string_teach[position];
        Intent intent = new Intent(SchTeActivity.this, ShowTeaDet_Activity.class);
        ...
        intent.putExtras(bundle);
        startActivity(intent);
    }
}
```

从上述代码中可以看到,当用户选择了课程和年级以后,就会执行一个异步任务去请求服务器,然后服务器会把客户端需要的数据返回到前端。这里网络请求的代码与第10章登录案例的代

码一样,只需要修改传递的参数即可。服务器端的 get_select_teach.php 的代码如下所示:

```php
<?php
    //获取教师表里的信息
    include('conn.php');
    //解决中文乱码问题
    $conn->query("SET NAMES 'UTF8'");
    $subject = $_POST["subject"];
    $grade = $_POST["grade"];
    $sql = "SELECT teach_name FROM teacher where grade = '$grade' and teach_sub = '$subject'";
    $result = $conn->query($sql);
    if ($result->num_rows > 0) {
        //输出每行数据
        while($row = $result->fetch_assoc()) {
            echo $row['teach_name'].",";
        }
    } else {
        echo "error";
    }
    $conn->close();
?>
```

其中引入的 conn.php 文件即是连接数据库的文件,前面已经多次讲解过,此处不再赘述。

完成了上述编码以后,即可查询教师。运行结果如图 11-17 所示。

以上就是查询教师的结果,当用户单击教师列表上的某一位教师时,就会显示该教师的详细信息,然后可以决定是否预约该教师。接下来讲解教师预约功能的实现。

图 11-17 查询教师结果

### 2. 教师预约

当用户选择教师时,会出现一个新的界面来显示教师的详细信息,这个新的界面对应的 Activity 为 ShowTeaDet_Activity。它对应的布局文件为 showteadetail.xml,该布局文件主要包含几个 TextView 组件和一个 Button 组件,这里对布局文件就不再介绍,具体可以查看项目下的文件。其 Activity 的功能是首先初始化各个组件,然后实现教师详细信息查询显示的功能以及实现教师预约的功能。该 Activity 的部分代码如下所示:

```java
public class ShowTeaDet_Activity extends Activity implements OnClickListener {
    private TextView sex,name,phonum,address,subject;
    private EditText exper;
    private ImageView icon;
    private Button order;
    private String subject_name;
    private String det_name;
    private Boolean isget = false;
```

```java
GetTeaDetail teaDetail = new GetTeaDetail();
Resrve_Teacher resrve = new Resrve_Teacher();
private String teach_phone,par_phone;

@Override
protected void onCreate(Bundle savedInstanceState) {
    super.onCreate(savedInstanceState);
    this.requestWindowFeature(Window.FEATURE_NO_TITLE);
    setContentView(R.layout.showteadetail);
    init();
    new AnotherTask().execute((Void[]) null);
}
//组件初始化
private void init(){
    Intent intent = this.getIntent();
    Bundle bundle = intent.getExtras();
    subject_name = bundle.getString("subject");
    det_name = bundle.getString("teach_name");
    par_phone = bundle.getString("par_phone");
    sex = (TextView)findViewById(R.id.det_sex);
    name = (TextView)findViewById(R.id.det_name);
    name.setText(det_name);
    phonum = (TextView)findViewById(R.id.det_phonum);
    address = (TextView)findViewById(R.id.det_address);
    subject = (TextView)findViewById(R.id.det_subject);
    subject.setText(subject_name);
    exper = (EditText)findViewById(R.id.det_exper);
    icon = (ImageView)findViewById(R.id.det_icon);
    order = (Button)findViewById(R.id.det_order);
    order.setOnClickListener(this);
}
@Override
public void onClick(View v) {
    //TODO Auto-generated method stub
    switch (v.getId()) {
    //预约教师
    case R.id.det_order:
        order.setEnabled(false);
        order.setBackgroundColor(Color.GRAY);
        Thread thread = new Thread(runnable);
        thread.start();
        break;
    default:
        break;
    }
}
//获取教师个人信息
private class AnotherTask extends AsyncTask< Void, Integer, Boolean > {
```

```java
        @Override
        protected Boolean doInBackground(Void... params) {
            //对 UI 组件的更新操作,是耗时的操作
            try {
                //TODO Auto-generated method stub
                String connecturl = "http://192.168.56.1/teacher_pro/teacher_detail.php";
                isget = teaDetail.getDetail(det_name, connecturl);

            } catch (Exception e) {
                e.printStackTrace();
            }
            return null;
        }
        @Override
        protected void onPostExecute(Boolean result) {
            if(isget){
                //获取具体信息数组
                String[] message = teaDetail.result.split(", ");
                System.out.println(message);
                sex.setText(message[0]);
                phonum.setText(message[1]);
                teach_phone = message[1];
                address.setText(message[2]);
                exper.setText("教学经历:" + message[3]);
            }
        }
    }

    //开启新的线程插入预约数据
    Runnable runnable = new Runnable() {
        @Override
        public void run() {
            String connecturl = "http://192.168.56.1/teacher_pro/par_reserve.php";
            Calendar now = Calendar.getInstance();
            String year = now.get(Calendar.YEAR) + "";
            String month = (now.get(Calendar.MONTH) + 1) + "";
            String day = now.get(Calendar.DAY_OF_MONTH) + "";
            Boolean flag = resrve.Reserve(par_phone, teach_phone, det_name,
                    subject_name, year + "-" + month + "-" + day, connecturl);
            if (flag) {
                Looper.prepare();
                Toast.makeText(ShowTeaDet_Activity.this, "预约成功,在消息中查看相关信息",
Toast.LENGTH_SHORT).show();
                Looper.loop();
            }
        }
    };
}
```

上述代码中的 teaDetail 所对应的网络请求类与查询教师的一样，修改参数即可。其查询教师详细信息对应的 teacher_detail.php 文件的代码如下所示：

```php
<?php
    //获取教师表里的信息
    include('conn.php');
    //解决中文乱码问题
    $conn->query("SET NAMES 'UTF8'");
    $name = $_POST["name"];
    $sql = "SELECT * FROM teacher where teach_name = '$name'";
    $result = $conn->query($sql);
    if ($result->num_rows > 0) {
        //输出每行数据
        while($row = $result->fetch_assoc()) {
            echo $row['teach_sex'].",". $row['teache_phonum'].",". $row['teach_address'].",
                ". $row['teach_exper'].",". $row['teach_icon'];
        }
    } else {
        echo "";
    }
    $conn->close();
?>
```

以上就是查询教师详细信息的 PHP 文件，当用户单击"预约教师"按钮后，会开启一个新的线程去实现教师的预约，即向教师预约表当中插入一条数据。其后台 par_reserve.php 执行文件的代码如下所示：

```php
<?php
    include 'conn.php';
    //解决中文乱码问题
    $conn->query("SET NAMES 'UTF8'");
    //获取预约的家长手机号和老师手机号、科目
    $par_phone = $_POST["par_phone"];
    $teach_phone = $_POST["teach_phone"];
    $name = $_POST["teach_name"];
    $subject = $_POST["subject"];
    $data = $_POST["data"];
    //在这里进行插入数据库操作
    $sql = "INSERT INTO reserve(parent_phone,teach_phone,teach_name,subject,data)
            VALUES ('$par_phone','$teach_phone','$name','$subject','$data')";
    if ($conn->query($sql) === TRUE) {
            echo "success";
    } else {
        echo "fail";
    }
$conn->close();
?>
```

以上就是预约教师功能所对应的后台服务器处理的详细代码,结果如图 11-18 所示。

图 11-18 教师预约结果

### 11.4.3 个人信息管理功能实现

个人信息管理包含的功能有个人订单的查询、钱包信息的查询、奖学券的查询以及个人学习计划的制订等。接下来主要讲解前 3 个功能的实现。

#### 1. 个人订单查询功能

个人订单查询主要是查询用户已经购买的书籍,包括已经收到货的和未收到货的,其对应的 Activity 是 OrderActivity,该 Activity 继承自 ListActivity,把查询的信息直接以列表的形式显示出来。该 Activity 的具体代码如下所示:

```java
public class OrderActivity extends ListActivity {
    private String phonenum = "";
    public Boolean order_flag = false;
    String content[] = {};
    List<Map<String, Object>> list;
    @Override
    protected void onCreate(Bundle savedInstanceState) {
        super.onCreate(savedInstanceState);
        this.requestWindowFeature(Window.FEATURE_NO_TITLE);
        Intent intent = this.getIntent();
        Bundle bundle = intent.getExtras();
        phonenum = bundle.getString("phone");
        SimpleAdapter adapter = new SimpleAdapter(OrderActivity.this, this.getData(),
            R.layout.order_layout, new String[]{"bookname", "booknum", "bookprice", "ztai"},
                new int[]{R.id.bookname, R.id.booknum, R.id.bookprice, R.id.ztai});
        setListAdapter(adapter);
```

```java
    }
    private List<? extends Map<String, Object>> getData() {
        System.out.println(content.length);
        list = new ArrayList<Map<String, Object>>();
        Thread thread = new Thread(runnable);
        thread.start();
        return list;
    }
    //开启新的线程用来进行后台订单数据获取
    Runnable runnable = new Runnable() {
        @Override
        public void run() {
            HttpMe httpMe = new HttpMe();
            //连接到服务器的地址
            String connectURL = "http://192.168.56.1/teacher_pro/order.php";
            order_flag = httpMe.getOrder(phonenum, connectURL);
            if (order_flag) {
                //取得返回的内容
                content = httpMe.result.split(", ");
                //添加列表内容
                Map<String, Object> map = new HashMap<String, Object>();
                System.out.println(content.length);
                for(int i = 0; i < content.length/4; i++){
                    map = new HashMap<String, Object>();
                    map.put("bookname", content[4 * i + 0]);
                    map.put("booknum", content[4 * i + 1]);
                    map.put("bookprice", content[4 * i + 2] + "元");
                    map.put("ztai", content[4 * i + 3]);
                    list.add(map);
                }
            }
        }
    };
}
```

从上述代码中可以看到，当用户单击"我的订单"图标时，就会跳转到该 Activity，然后该 Activity 通过重新开启一个线程的方式来进行网络请求操作，其后台处理 order.php 文件的具体代码如下所示：

```php
<?php
    //获取订单表里的信息
    include('conn.php');
    //解决中文乱码问题
    $conn->query("SET NAMES 'UTF8'");
    $phone = $_POST["phone"];
    $sql = "SELECT bookname, count, price, state FROM par_order where par_phone = '$phone'";
    $result = $conn->query($sql);
```

```php
        if ( $ result -> num_rows > 0) {
            //输出每行数据
            while( $ row = $ result -> fetch_assoc()) {
                echo $ row['bookname'].",". $ row['count'].",". $ row['price'].",".
                $ row['state'].",";
            }
        } else {
            echo "";
        }
        $ conn -> close();
    ?>
```

当完成上述编码以后，即可运行程序。程序的运行结果如图11-19所示。

**2. 钱包信息查询功能**

当用户单击"我的钱包"图标时，既可查看个人电子钱包信息，包含了总额、余额、积分等信息。其所对应的Activity为MoneyActivity，该Activity所加载的布局文件为money_layout，前面已经讲过。接下来讲解该Activity的代码，如下所示：

图 11-19 订单查询结果

```java
package com.example.teacher_pro.me;
import com.example.teacher_pro.R;
public class MoneyActivity extends Activity {
    private String phonenum = "";
    private Boolean isSucceed = false;
    HttpMe httpMe = new HttpMe();
    //用来存储钱包信息
    private String[] message = {};
    //获取布局文件的组件
    private TextView zong, yue, jifen;
    @Override
    protected void onCreate(Bundle savedInstanceState) {
        super.onCreate(savedInstanceState);
        this.requestWindowFeature(Window.FEATURE_NO_TITLE);
        setContentView(R.layout.money_layout);
        Intent intent = this.getIntent();
        Bundle bundle = intent.getExtras();
        phonenum = bundle.getString("phone");
        init();
        new AnotherTask().execute((Void[]) null);
    }
    //组件初始化
    private void init(){
        zong = (TextView)findViewById(R.id.zong);
        yue = (TextView)findViewById(R.id.yue);
```

```
            jifen = (TextView)findViewById(R.id.jifen);
    }
    //获取钱包信息
    private class AnotherTask extends AsyncTask< Void, Integer, Boolean > {
            @Override
            protected Boolean doInBackground(Void... params) {
                //对 UI 组件的更新操作,是耗时的操作
                try {
                    //连接到服务器的地址
                    String connectURL = "http://192.168.56.1/teacher_pro/money.php";
                    //输入用户名密码和连接地址
                    isSucceed = httpMe.getOrder(phonenum, connectURL);
                } catch (Exception e) {
                    e.printStackTrace();
                }
                return null;
            }
            @Override
    protected void onPostExecute(Boolean result) {
            if (isSucceed) {
                message = httpMe.result.split(",");
                zong.setText("我的总额:" + message[0]);
                yue.setText("我的余额:" + message[1]);
                jifen.setText("我的积分:" + message[2]);
            }
        }
    }
}
```

从上述代码中可以看到,该 Activity 首先对布局文件中的各个组件进行了初始化,然后从加粗的那一句代码开始请求服务器来执行钱包信息的查询。首先重写了一个内部类,然后让它继承自异步类,重写其中的方法,然后实现钱包信息的查询以及 UI 界面的更新。

服务器处理 money.php 文件的具体代码如下所示:

```
<?php
    //获取订单表里的信息
    include('conn.php');
    //解决中文乱码问题
    $conn->query("SET NAMES 'UTF8'");
    $phone = $_POST["phone"];
    $sql = "SELECT par_balance, yue, jifen FROM par_money where phonenum = '$phone'";
    $result = $conn->query($sql);
    if ($result->num_rows > 0) {
        //输出每行数据
        while($row = $result->fetch_assoc()) {
            echo $row['par_balance'].",".$row['yue'].",".$row['jifen'];
        }
```

```
        } else {
            echo "";
        }
        $conn->close();
?>
```

以上代码执行的结果如图 11-20 所示。

图 11-20　钱包信息查询结果

### 3. 奖学券查询功能

当用户单击"奖学券"图标时，就会去请求服务器将奖学券的信息返回到前端显示出来。其对应的 Activity 为 RewardActivity，该 Activity 继承自 ListActivity，把查询到的信息以列表的形式显示出来。其具体代码如下所示：

```java
public class RewardActivity extends ListActivity {
    private String phonenum = "";
    public Boolean order_flag = false;
    String content[] = {};
    List<Map<String, Object>> list;
    @Override
    protected void onCreate(Bundle savedInstanceState) {
        super.onCreate(savedInstanceState);
        this.requestWindowFeature(Window.FEATURE_NO_TITLE);
        Intent intent = this.getIntent();
        Bundle bundle = intent.getExtras();
        phonenum = bundle.getString("phone");
        SimpleAdapter adapter = new SimpleAdapter(RewardActivity.this,
                this.getData(), R.layout.reward_layout, new String[]{"jine", "endtime"},
                new int[]{R.id.jine, R.id.endtime});
        setListAdapter(adapter);
    }
    private List<? extends Map<String, Object>> getData() {
        System.out.println(content.length);
        list = new ArrayList<Map<String, Object>>();
        Thread thread = new Thread(runnable);
        thread.start();
        return list;
    }
```

```java
//开启新的线程用来进行后台订单数据获取
Runnable runnable = new Runnable() {
    @Override
    public void run() {
        HttpMe httpMe = new HttpMe();
        //连接到服务器的地址
        String connectURL = "http://192.168.56.1/teacher_pro/reward.php";
        order_flag = httpMe.getOrder(phonenum, connectURL);
        if (order_flag) {
            //取得返回的内容
            content = httpMe.result.split(",");
            //添加列表内容
            Map<String, Object> map = new HashMap<String, Object>();
            System.out.println(content.length);
            for(int i = 0;i < content.length/2;i++){
                map = new HashMap<String, Object>();
                map.put("jine", content[2 * i + 0]);
                map.put("endtime", content[2 * i + 1]);
                list.add(map);
            }
        }
    }
};
```

从上面的加粗代码中可以看到,在涉及网络请求等一些耗时性的操作时,需要开启新的线程或者使用异步任务来执行。其中 HttpMe 实现了发送网络请求以及获取返回的数据的功能。HttpMe 的具体代码如下所示:

```java
public class HttpMe {
    //获取返回的信息
    public String result = "";                  //用来取得返回的 String
    public boolean getOrder(String phonenum, String connectUrl) {
        boolean isLoginSucceed = false;
        HttpClient httpClient = new DefaultHttpClient();
        //发送 POST 请求
        HttpPost httpRequest = new HttpPost(connectUrl);
        //POST 运作传送变数必须用 NameValuePair[]阵列存储
        List<NameValuePair> params = new ArrayList<NameValuePair>();
        params.add(new BasicNameValuePair("phone", phonenum));
        try {
            //发出 HTTP 请求
            httpRequest.setEntity(new UrlEncodedFormEntity(params));
            //取得 HTTP Response
            HttpResponse httpResponse = httpClient.execute(httpRequest);
            result = EntityUtils.toString(httpResponse.getEntity(),"UTF-8");
            System.out.println(result);
```

```
            } catch (Exception c) {
                e.printStackTrace();
            }
            //判断返回的数据是否为 PHP 中成功登录时输出的
            if (!result.equals("")) {
                isLoginSucceed = true;
            }
            return isLoginSucceed;
        }
    }
```

上述代码的功能就是向服务器发送请求,并且取得返回的数据,其后台服务器负责处理请求的 reward.php 的文件代码如下所示:

```
<?php
    //获取奖学券表里的信息
    include('conn.php');
    //解决中文乱码问题
    $conn->query("SET NAMES 'UTF8'");
    $phone = $_POST["phone"];
    $sql = "SELECT acount,data FROM par_reward where phonenum = '$phone'";
    $result = $conn->query($sql);
    if ($result->num_rows > 0) {
        //输出每行数据
        while($row = $result->fetch_assoc()) {
            echo $row['acount'].",". $row['data'].",";
        }
    } else {
        echo "";
    }
    $conn->close();
?>
```

以上就是查询奖学券的详细代码,其执行的结果如图 11-21 所示。

图 11-21 奖学券查询结果

### 11.4.4 预约的教师查询功能

预约的教师查询功能是指当用户单击消息界面上的"预约老师消息"时,会弹出已经预约过的教师信息,其中有的教师可能预约了不止一次,会根据预约的时间来区分。该功能对应的 Activity 为 order_teachActivity。该 Activity 首先对布局文件中的组件进行初始化,然后通过异步任务来发送网络请求并获取数据在前端显示出来。该 Activity 的代码如下所示:

```java
public class order_teachActivity extends Activity implements OnItemClickListener {
    private ListView teach_list;
    private String[] string = new String[20];
    private String phone_num;
    private HttpMe http_me = new HttpMe();
    private Boolean isSucceed;
    private String[] teach_name = new String[2];
    @Override
    protected void onCreate(Bundle savedInstanceState) {
        super.onCreate(savedInstanceState);
        this.requestWindowFeature(Window.FEATURE_NO_TITLE);
        setContentView(R.layout.mes_order_teach);
        init();
        new Anothertask().execute((Void[]) null);
    }
    //组件初始化方法
    private void init(){
        //获取登录的手机号
        Intent intent = this.getIntent();
        Bundle bundle = intent.getExtras();
        phone_num = bundle.getString("phone");
        teach_list = (ListView)findViewById(R.id.order_teach_list);
        this.registerForContextMenu(teach_list);
        teach_list.setOnItemClickListener(this);
    }
    //异步任务获取预约教师信息
    private class Anothertask extends AsyncTask< Void, Integer, Boolean >{
        @Override
        protected Boolean doInBackground(Void... params) {
            //对 UI 组件的更新操作,是耗时的操作
            try {
                //连接到服务器的地址
                String connectURL = "http://192.168.56.1/teacher_pro/get_order_teach.php";
                //输入用户名、密码和连接地址
                isSucceed = http_me.getOrder(phone_num, connectURL);
            } catch (Exception e) {
                e.printStackTrace();
            }
            return null;
        }
        @Override
        protected void onPostExecute(Boolean result) {
```

```java
            //TODO Auto-generated method stub
            if (isSucceed) {
                string = http_me.result.split(",");
                teach_list.setAdapter(newArrayAdapter<String>(order_teachActivity.this,
                                      R.layout.array_adapt, string));
            }
        }
    }
    @Override
    public void onItemClick(AdapterView<?> parent, View view, int position, long id) {
        //TODO Auto-generated method stub
        teach_name = string[position].split(" ");
        Intent intent = new Intent(order_teachActivity.this, ShowTeaDet_Activity.class);
        Bundle bundle = new Bundle();
        bundle.putString("teach_name", teach_name[1]);
        intent.putExtras(bundle);
        bundle.putString("subject", teach_name[0]);
        bundle.putString("par_phone", phone_num);
        startActivity(intent);
    }
}
```

上述异步任务发送网络请求的地址为 get_order_teach.php，该文件负责处理网络请求，并将查询到的用户预约教师的信息返回到前端。该文件的具体代码如下所示：

```php
<?php
    //获取预约教师表里的信息
    include('conn.php');
    //解决中文乱码问题
    $conn->query("SET NAMES 'UTF8'");
    $phone = $_POST["phone"];
    $sql = "SELECT subject,teach_name FROM reserve where paret_phone = '$phone'";
    $result = $conn->query($sql);
    if ($result->num_rows > 0) {
        //输出每行数据
        while($row = $result->fetch_assoc()) {
            echo $row['subject']." ".$row['teach_name'].",";
        }
    } else {
        echo "";
    }
    $conn->close();
?>
```

以上就是用户已经预约的教师查询功能的代码。该代码执行的结果如图 11-22 所示。

图 11-22 已预约教师查询结果

## 11.4.5 设置功能的实现

设置功能主要包含新消息提醒的设置、聊天的设置、账号安全的设置以及应用程序的退出功能。该部分功能尚未实现，布局文件和 Activity 已经写好了，只需要调用系统一些自带的功能，如系统的铃声、来消息时的震动等。该页面对应的布局文件为 me_setting.xml，其代码如下所示：

```xml
<?xml version = "1.0" encoding = "utf-8"?>
<LinearLayout
    xmlns:android = "http://schemas.android.com/apk/res/android"
    android:orientation = "vertical"
    android:background = "#E8E8E8"
    android:layout_width = "match_parent"
    android:layout_height = "match_parent">
    <LinearLayout
        android:layout_width = "match_parent"
        android:layout_height = "51dp"
        android:background = "#11cd6e"
        android:orientation = "horizontal">
        <!-- 返回箭头 -->
        <ImageView
            android:layout_width = "1dp"
            android:layout_weight = "0.5"
            android:layout_height = "30dp"
            android:layout_marginTop = "10dp"
            android:contentDescription = "@string/app_name"
            android:src = "@drawable/arrowleft" />
        <!-- 标题 -->
        <LinearLayout
            android:layout_width = "0dp"
            android:layout_weight = "5"
            android:layout_height = "50dp"
            android:orientation = "vertical">
            <TextView
                android:layout_width = "wrap_content"
                android:layout_gravity = "center"
                android:layout_height = "wrap_content"
                android:textSize = "20sp"
                android:textColor = "#FFFFFF"
```

```xml
                    android:layout_marginTop = "14dp"
                    android:layout_marginRight = "17dp"
                    android:text = "设置"/>
            </LinearLayout>
        </LinearLayout>
    ... <!-- 与上面代码一致,不再讲解 -->
            < LinearLayout
                android:layout_width = "fill_parent"
                android:layout_height = "45dp"
                android:layout_marginLeft = "10dp"
                android:orientation = "horizontal" >
                < ImageView
                    android:layout_width = "0dp"
                    android:layout_weight = "0.5"
                    android:layout_height = "wrap_content"
                    android:layout_gravity = "center_vertical"
                    android:contentDescription = "@string/app_name"
                    android:src = "@drawable/kecheng" />
                < TextView
                    android:layout_width = "0dp"
                    android:layout_weight = "5"
                    android:layout_height = "wrap_content"
                    android:layout_gravity = "center"
                    android:layout_marginLeft = "15dp"
                    android:text = "退出"
                    android:textSize = "16sp" />
                < ImageView
                    android:layout_width = "0dp"
                    android:layout_weight = "1"
                    android:layout_height = "wrap_content"
                    android:layout_gravity = "center_vertical"
                    android:contentDescription = "@string/app_name"
                    android:src = "@drawable/arrowright" />
            </LinearLayout>
        </LinearLayout>
</LinearLayout>
```

以上就是设置界面所使用到的布局文件代码,学习者可以使用这个布局文件,自定义一个Activity来实现设置界面的各个功能。

至此,本应用程序所使用到的数据库、界面的布局设计、功能的实现已经全部讲解完了。最重要的一点是学会使用 HttpClient 去发送网络请求,学会参数的传递、服务器参数的读取以及服务器与数据库之间的连接交互等。掌握了本章的开发,就可以进行有关网络方面的应用程序开发了。

## 11.5 应用发布

在完成了"倾心家教"App 的开发以后,就可以在应用市场进行发布。本 App 选择在百度移动应用上发布。具体的发布步骤如下。

1. 百度账户注册以及实名制认证

首先登录百度账号,然后在 http://app.baidu.com/user/register 上完成实名制注册。操作如图 11-23 所示。

图 11-23 实名制认证

填写资料完成后,等待百度官方审核,审核一般需要 3~5 天。审核完成后,开发者提交个人资料,包括开发者姓名、联系地址、手机号等,然后就可以使用百度云服务。提交以后的结果如图 11-24 所示。

图 11-24 开发者提交资料

2. 进入管理平台发布应用

在实名制认证以及开发者资料提交完成以后,接着就可以进入管理平台发布应用。在管理平台上选择发布应用的操作如图 11-25 和图 11-26 所示。

图 11-25　创建应用

图 11-26　选择 App 应用

在选择完 App 应用程序以后,会让开发者填写应用程序的名称、图标以及验证码,然后提交以后就可以通过打包 APK、上传应用程序信息,完成以后应用程序就可以发布在百度市场上。具体操作如图 11-27 所示。

图 11-27　上传 App 应用

## 11.6　本章小结

本章主要通过一个家教 App 案例来讲解 Android＋PHP＋MySQL 的结合。首先讲解了家教类 App 当前的市场,确定项目的主题,通过用例图来分析该应用程序应当具备哪些功能;接着又把该应用程序需要使用的数据库建立起来;最后进行功能的实现。整个过程是进行有关网络应用程序开发必须要掌握的,只有这样,才能更好、更快地开发应用程序。

# 第 12 章

# "橙一色"App 功能优化及完善

## 12.1 应用分析

前面章节中已经完成了"橙一色"App 的登录、注册、首页界面的设计以及个人信息的展示等功能,但是使用的数据都是一些模拟数据,如果需要改动的话,就需要对程序的代码进行修改更新。如何在不改变 Android 端界面的前提下去修改界面上的信息是本章要解决的问题。

本章介绍如何通过更改数据库数据的方式来动态更新 Android 端界面。数据库的更新的方法包括界面化的操作和 SQL 语句两种更新方式。本章采用界面化的操作来更新。Android 与 MySQL 数据的连接,不再选取第 11 章中用到的 PHP,而是使用目前流行的 Java 开源框架 Spring Boot,结合 MyBatis+MySQL 进行开发实现,这样可以让读者对 Java 方面的知识有更多的了解。

## 12.2 App 应用技术

视频讲解

"橙一色"App 应用的技术包括 Android、Spring Boot、MyBatis、MySQL、Vue+Element 以及 JSON 解析库 Fastjson。各项技术的应用场景如下所述。

- App 前端:使用 Android 中的布局进行设计,用于数据的展示。
- 后台管理前端:使用 Vue+Element 设计实现,主要是用来提供一个数据库增加、删除、查询和修改的操作界面,方便管理人员进行操作。
- 服务器端:后台通过 Spring Boot+MyBatis+MySQL 进行数据的操作,数据以 JSON 的格式进行交互。

本章简要介绍这些技术的基础知识,后续大家可以进行深入的学习。

### 12.2.1 Spring Boot 框架

在学习 Spring Boot 之前,先来了解一下 Spring。Spring 是一个开源框架,它是一个轻量级的 Java 开发框架,它的出现是为了解决企业级应用开发的复杂性问题,使用 Spring 可以让简单的

JavaBean 实现之前只有 EJB 才能完成的事情。但是它不仅仅局限于服务器端开发,任何 Java 应用都能在简单性、可测试性和松耦性等方面从 Spring 获益。Spring 独立于各种应用服务器,基于 Spring 框架的应用,可以真正实现 Write Once,Run Anywhere(一次编写,到处运行)。

为了降低 Java 开发的复杂性,Spring 采用了以下 4 种关键策略:

(1) 基于 POJO(Plain Old Java Object,普通老式 Java 对象)的轻量级和最小侵入性编程——Spring 不会强迫用户实现 Spring 规范的接口或继承 Spring 规范的类,而是通过依赖注入来装载自身的 API,尽量避免因自身的 API 而弄乱用户的应用代码。一个类或许会使用 Spring 注解,但它依旧是 POJO。

(2) 通过依赖注入(Dependency Injection,DI)和面向接口实现松耦合——Spring 的 DI 机制降低了业务对象替换的复杂性,提高了组件之间的解耦。

(3) 基于切面(Aspect Oriented Program,AOP)和惯例进行声明式编程——Spring 的 AOP 支持允许将一些通用任务如安全、事务和日志等进行集中式管理,从而提供了更好的复用。

(4) 通过切面和模板减少样板式代码——Spring 的 ORM 和 DAO 提供了与第三方持久层框架的良好整合,并简化了底层的数据库访问。

Spring Boot 是所有基于 Spring 开发的项目的起点,它的设计是为了让用户尽快搭建一个 Spring 应用,并且尽可能减少配置文件。Spring Boot 的 4 个主要特性:①Spring Boot Starter,它将常用的依赖分组进行了整合,将其合并到一个依赖中;②利用 Spring 4 对条件化配置的支持,合理地推测应用所需的 Bean 并自动配置它们;③命令行接口;④Actuatir,它为 Spring Boot 应用的所有特性构建一个小型的应用程序。

Spring Boot 的官方网站地址为 https://spring.io/projects/spring-boot/,如图 12-1 所示。

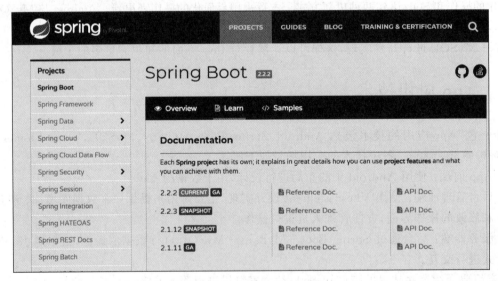

图 12-1  Spring Boot 官方网站

搭建一个 Spring Boot 项目主要有两种方式:

(1) 利用 Spring 官方网站下的 Spring Initializr 来搭建,如图 12-2 所示。

(2) 利用 IntelliJ IDEA 开发工具搭建项目结构。

图 12-2　Spring Initializr 界面

本章使用 IDEA 开发工具来搭建 Spring Boot 项目。首先需要下载 IDEA 开发工具，官方网站地址为 http://www.jetbrains.com/，如图 12-3 所示。

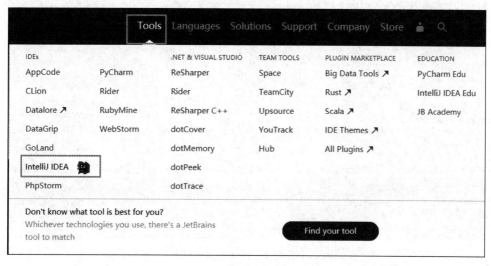

图 12-3　IDEA 下载界面

单击 IntelliJ IDEA，即可完成开发工具的下载。接下来使用 IDEA 创建一个 Spring Boot 项目。打开 IDEA，选择项目类型为 Spring Initializr，然后直接单击 Next 按钮，如图 12-4 所示。

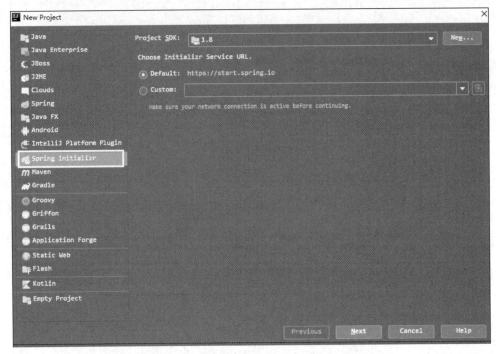

图 12-4　利用 IDEA 创建 Spring Boot 项目

自定义新创建项目的名称、包名等信息，然后一直单击 Next 按钮即可创建一个 Spring Boot 项目。具体操作界面如图 12-5 和图 12-6 所示。

图 12-5　操作步骤 1

图 12-6　操作步骤 2

在图 12-6 中单击 Finish 按钮，完成一个 Spring Boot 项目的创建。Spring Boot 有内置的 Tomcat，默认端口号为 8080，所以不需要进行额外的配置即可启动项目。单击 IDEA 右上方的 Run 按钮即可运行，如图 12-7 和图 12-8 所示。

图 12-7　运行项目

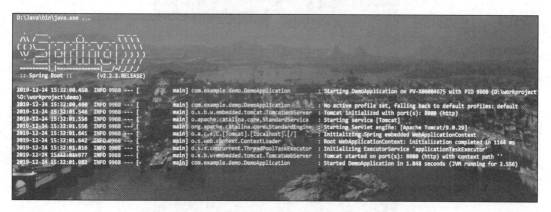

图 12-8　运行结果

至此，一个 Spring Boot 项目就创建完成了，如果需要写一些接口就可以在项目里面进行添加。如针对一般的数据库的增加、删除、查询和修改操作，可以结合 MyBatis 一起开发。

## 12.2.2　MyBatis 持久层框架

MyBatis 是一款优秀的持久层框架，它支持定制化 SQL、存储过程以及高级映射等。MyBatis

避免了几乎所有的 JDBC 代码和手动设置参数以及获取结果集。MyBatis 可以使用简单的 XML 或注解来配置和映射原生类型、接口和 Java 的 POJO 为数据库中的记录。

MyBatis 的中文官方网站地址为 https://mybatis.org/mybatis-3/zh/index.html，通过学习官方网站上的知识就可快速学会使用 MyBatis 的方法。通常都是将 MyBatis 和 Spring Boot 结合使用，所以首先需要在 Spring Boot 项目中添加依赖。具体内容如下所示：

```
<dependency>
    <groupId>org.mybatis</groupId>
    <artifactId>mybatis</artifactId>
    <version>x.x.x</version>
</dependency>
```

接着就可以使用 MyBatis 了。使用 MyBatis 时通常是定义一个接口，并且加上 @Mapper 注解；然后在 resource 下创建一个 mapper 文件夹，在文件夹下写一个与接口对应着的 mapper 文件，文件名与接口名保持一致，防止出现找不到的问题。具体案例的项目结构如图 12-9 所示。

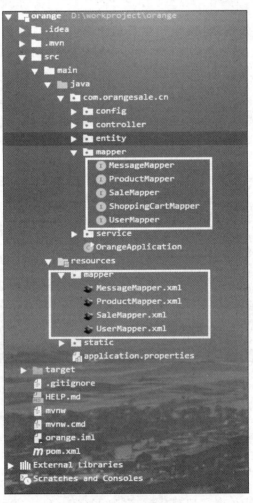

图 12-9　项目结构图

图 12-9 是一个完整的项目结构图,其中 OrangeApplication 是 Spring Boot 项目的入口类,即启动类;config 包下是项目的一些配置类;controller 即是服务器端的接口,定义接口的地址,请求方法如 Get();entity 是实体类,通常是与数据库中表的字段相对应,service 是连接 controller 与 mapper 之间的类,通常 controller 不会直接调用 mapper,而是通过 service 调用 mapper,将返回的结果进行处理之后再返回给 controller。这样的设计使得项目的结构层次分明,便于项目管理。

下面重点介绍 Mapper 接口与文件之间的映射是如何操作的。首先看一下 Mapper 接口的内容:

```java
@Mapper
public interface UserMapper {
    /** 用户登录验证
     * @param orangeUser
     * @return
     */
    OrangeUser userLogin(OrangeUser orangeUser);
    /** 用户注册
     * @param orangeUser
     * @return
     */
    Integer userRegister(OrangeUser orangeUser);
}
```

上面接口中主要定义了两个方法——用户登录验证及用户注册,对应着数据库的查找与插入操作。接着看一下对应的 mapper 文件的内容:

```xml
<?xml version="1.0" encoding="utf-8"?>
<!DOCTYPE mapper PUBLIC "-//mybatis.org//DTD Mapper 3.0//EN"
        "http://mybatis.org/dtd/mybatis-3-mapper.dtd">
<mapper namespace="com.OrangeSale.cn.mapper.UserMapper">
    <select id="userLogin" resultType="com.OrangeSale.cn.entity.OrangeUser">
    select * from orange_user where username = #{username} and password = #{password}
    </select>
    <insert id="userRegister">
    insert into orange_user(username, password, sex, city) values(#{username}, #{password},
                    #{sex}, #{city})
    </insert>
</mapper>
```

从以上加粗部分的代码可以看到,namespace 需要是对应的类名或接口名,id 分别是用户登录及注册的方法名。关于 mapper 文件里面的写法大家可以自行去官方网站上学习掌握。

## 12.2.3 Fastjson 解析库

目前来说,前后端通常是以 JSON 的数据格式进行交互的,JSON 数据格式的封装与解析是经常要使用到的,因此目前主流的 Fastjson 解析库是比较常用的。

阿里官方给的定义是：Fastjson是阿里巴巴的开源JSON解析库，它可以解析JSON格式的字符串，支持将JavaBean序列化为JSON字符串，也可以从JSON字符串反序列化到JavaBean。Fastjson具有速度快、使用广泛、测试完备、使用简单以及功能完备等优点，所以使用起来易于上手以及操作安全。

Fastjson的入口类是com.alibaba.fastjson.JSON，它的两个主要的API是JSON.toJSONString和JSON.parseObject。具体内容如下所示：

```java
package com.alibaba.fastjson;
public abstract class JSON {
    //将Java对象转换为JSON字符串
    public static final String toJSONString(Object object);
    //将JSON字符串转换为Java对象
    public static final <T> T parseObject(String text, Class<T> clazz, Feature... features);
}
```

序列化：

```java
String jsonString = JSON.toJSONString(obj);
```

反序列化：

```java
VO vo = JSON.parseObject("...",VO.class);
```

泛型反序列化：

```java
import com.alibaba.fastjson.TypeReference;
List<VO> list = JSON.parseObject("...",new TypeReference<List<VO>>() {});
```

这种方式，通常可以将JSON字符串转为一个对象的List集合，从而可以快速地解析JSON字符串进行应用。

下面可以通过一个班级学生的案例来说明Fastjson是如何使用的。首先创建一个班级Grade对象，具体内容如下所示：

```java
public class Grade {
    private Long id;
    private String name;
    private List<Student> users = new ArrayList<Student>();
    //省略 setter、getter
    public void addStudent(Student student) {
        users.add(student);
    }
    @Override
    public String toString() {
        return "Grade{" +
                "id=" + id +
```

```
        ", name = '" + name + '\'' +
        ", users = " + users +
        '}';
    }
}
```

接着创建一个 Student 对象，具有 id 和 name 的属性，具体内容如下所示：

```java
public class Student {
    private Long id;
    private String name;
    //省略 setter、getter
    @Override
    public String toString() {
        return "Student{" +
                "id = " + id +
                ", name = '" + name + '\'' +
                '}';
    }
}
```

最后写一个测试的 main() 函数，测试 JavaBean 对象与 JSON 字符串之间的互相转换，具体内容如下所示：

```java
public class Test {
    public static void main(String[] args) {
        Grade group = new Grade();
        group.setId(0L);
        group.setName("admin");
        Student student = new Student();
        student.setId(2L);
        student.setName("guest");
        Student rootUser = new Student();
        rootUser.setId(3L);
        rootUser.setName("root");
        group.addStudent(student);
        group.addStudent(rootUser);
        String jsonString = JSON.toJSONString(group);
        System.out.println("JSON 字符串:" + jsonString);
        Grade grade = JSON.parseObject(jsonString, Grade.class);
        System.out.println("JavaBean 对象:" + grade);
    }
}
```

程序的最终输出结果如下所示：

```
JSON 字符串:
{"id":0,"name":"admin","users":[{"id":2,"name":"guest"},{"id":3,"name":"root"}]}
```

```
JavaBean 对象:
Grade{id = 0, name = 'admin', users = [Student{id = 2, name = 'guest'}, Student{id = 3, name = 'root'}]}
```

Fastjson 还有很多其他属性,如对象空值的输出、日期的处理、定制序列化等大家可以在课下进行深入的学习。

接下来介绍 Android 中 Fastjson 的使用,关于解析与封装与上述代码一致,主要是需要在 build.gradle 中添加依赖,具体内容如下所示:

```
implementation group: 'com.alibaba', name: 'fastjson', version: '1.2.12'
```

## 12.3 "橙一色"App 应用功能界面设计

视频讲解

在前面章节的项目案例中已经完成了登录、注册、首页以及个人信息界面的设计,接下来设计商品界面以及购物车的界面。

### 12.3.1 商品界面

商品界面主要包括 3 部分:商品标题与返回箭头、一个商品查询条件的下拉列表框、商品的展示。具体内容如下所示:

```
<?xml version = "1.0" encoding = "utf-8"?>
<LinearLayout xmlns:android = "http://schemas.android.com/apk/res/android"
    android:layout_width = "match_parent"
    android:layout_height = "match_parent"
    android:orientation = "vertical">
    <include layout = "@layout/content_product_title" />
    <Spinner
        android:id = "@ + id/product_select_condition"
        android:layout_width = "match_parent"
        android:paddingHorizontal = "1dp"
        android:layout_height = "wrap_content" />
    <GridView
        android:id = "@ + id/product_list"
        android:layout_width = "match_parent"
        android:layout_height = "wrap_content"
        android:background = "#E8E8E8"
        android:numColumns = "2"/>
</LinearLayout>
```

content_product_title 是商品的标题与返回箭头布局文件,是一个水平的线性布局,里边包含了 ImageView 和 TextView 两个组件,如图 12-10 中上面一行的标题。这里不再列出该文件的代码。

完成了商品展示界面布局文件后,需要为 Spinner 和 GridView 填充数据,这些都是在商品对应的 Fragment：ProductFragment 里边完成的。这些将在功能实现里进行讲解。商品展示的最终布局如图 12-10 所示。

图 12-10　商品展示界面

## 12.3.2　购物车界面

购物车分为两种情况：购物车有商品时；购物车无商品时。购物车有商品时，可以参考商品展示界面，使用一个标题和返回箭头布局加上一个 GridView 组件，其中 GridView 每行只显示一个商品，最后再加上一个商品编辑的功能即可。本节主要介绍的是商品列表无商品时的布局，当无商品时，单击"去逛逛"按钮跳转到商品界面浏览商品。具体内容如下所示：

```
<?xml version = "1.0" encoding = "utf-8"?>
< LinearLayout xmlns:android = "http://schemas.android.com/apk/res/android"
    android:layout_width = "match_parent"
    android:layout_height = "match_parent"
    android:background = "#F2F2F2"
    android:orientation = "vertical">
```

```xml
<include layout = "@layout/shoppingcart_title" />
<!-- 无商品时的布局 -->
<LinearLayout
    <!-- 省略高宽的定义 -->
    android:orientation = "vertical">
    <ImageView
        <!-- 省略高宽的定义 -->
        android:src = "@drawable/cart" />
    <TextView
        <!-- 省略高宽的定义 -->
        android:text = "购物车是空的~"
        android:textColor = "#B5B5B5"
        android:textSize = "20sp" />
    <Button
        android:id = "@ + id/random_search"
        <!-- 省略高宽的定义 -->
        android:text = "去逛逛"
        android:textColor = "#fff"
        android:textSize = "18sp" />
</LinearLayout>
</LinearLayout>
```

按钮的单击事件将在购物车对应的 Fragment：ShoppingCartFragment 中实现，购物车没有商品时的布局界面如图 12-11 所示。

图 12-11　购物车无商品时的布局界面

## 12.4 数据库创建

视频讲解

本节只实现"橙一色"App的部分功能：用户的登录、注册、首页商品的展示、个人信息的展示、商品界面的展示。其他的功能读者可以参考之前的功能实现方式进行完善。本地创建一个名为orange的数据库表，具体过程介绍如下。

### 1. 用户表 orange_user 的设计

用户表包含了id、用户名、密码、性别及城市信息。用户表的字段设计如图12-12所示。

| 名 | 类型 | 长度 | 小数点 | 不是 null | 虚拟 | 键 | 注释 |
|---|---|---|---|---|---|---|---|
| id | int | 11 | 0 | ☑ | ☐ | 🔑1 | 主键 |
| username | varchar | 32 | 0 | ☑ | ☐ | | |
| password | varchar | 32 | 0 | ☑ | ☐ | | |
| sex | varchar | 10 | 0 | ☑ | ☐ | | |
| city | varchar | 128 | 0 | ☑ | ☐ | | |

图 12-12　orange_user 表设计

### 2. orange_product 商品表的设计

商品表字段具体如图12-13所示。

| 名 | 类型 | 长度 | 小数点 | 不是 null | 虚拟 | 键 | 注释 |
|---|---|---|---|---|---|---|---|
| id | int | 10 | 0 | ☑ | ☐ | 🔑1 | 自增主键 |
| name | varchar | 64 | 0 | ☑ | ☐ | | 商品名字 |
| img_url | varchar | 255 | 0 | ☑ | ☐ | | 商品图片地址 |
| price | decimal | 10 | 2 | ☑ | ☐ | | 商品价格 |
| add_time | datetime | 0 | 0 | ☑ | ☐ | | 商品上市时间 |

图 12-13　orange_product 表设计

### 3. orange_sale 商品销量表设计

商品销量表主字段具体如图12-14所示。

| 名 | 类型 | 长度 | 小数点 | 不是 null | 虚拟 | 键 | 注释 |
|---|---|---|---|---|---|---|---|
| id | int | 10 | 0 | ☑ | ☐ | 🔑1 | 自增主键 |
| product_id | int | 11 | 0 | ☑ | ☐ | | 商品id |
| sales_num | int | 11 | 0 | ☑ | ☐ | | 商品销量 |

图 12-14　orange_sale 表设计

### 4. orange_shoppingcart 表设计

购物车表字段具体内容如图12-15所示。

### 5. orange_message 表设计

消息表的设计是为了完成向"橙一色"App推送消息的功能，具体字段如图12-16所示。

| 名 | 类型 | 长度 | 小数点 | 不是 null | 虚拟 | 键 | 注释 |
|---|---|---|---|---|---|---|---|
| id | int | 10 | 0 | ☑ | ☐ | 🔑1 | 自增主键 |
| user_id | int | 11 | 0 | ☑ | ☐ | | 用户id |
| product_id | int | 11 | 0 | ☑ | ☐ | | 商品id |
| num | int | 11 | 0 | ☑ | ☐ | | 数量 |

图 12-15　orange_shoppingcart 表设计

| 名 | 类型 | 长度 | 小数点 | 不是 null | 虚拟 | 键 | 注释 |
|---|---|---|---|---|---|---|---|
| id | int | 10 | 0 | ☑ | ☐ | 🔑1 | 自增主键 |
| content | varchar | 255 | 0 | ☑ | ☐ | | 消息内容 |
| send_time | datetime | 0 | 0 | ☑ | ☐ | | 发送时间 |
| used | enum | 0 | 0 | ☑ | ☐ | | 是否已读 |

图 12-16　orange_message 表设计

视频讲解

## 12.5　Web 前端界面设计

Web 端界面设计的目的是不通过手动的方式操作数据库去进行数据库的增加、删除、查找和修改，而是通过界面操作去执行，本节使用 Vue＋ElementUI 的方式去实现。这里就不对技术进行介绍了，仅展示操作的界面。后续读者可以通过对这种技术进行学习或者通过其他方式进行实现，如 JSP 等实现。

### 1. 首页设计

首页主要分为 3 部分：左边的导航、上方的标题栏以及右边的内容展示。导航栏的功能有用户管理、商品管理、产品销量、消息推送以及设置等。本节按照教学案例的方式实现了部分功能，完整代码附在本书的电子资源当中，读者可以参考已有的功能代码进行完善。

首页的设计如图 12-17 所示。

图 12-17　首页设计图

## 2. 商品添加页面设计

商品添加页面主要包含了商品名称、上市时间、商品价格、商品图片等。实现如图 12-18 所示。

图 12-18　商品添加页面

## 3. 商品查询页面设计

商品查询页面目前设计的是根据输入的商品名字去模糊查询商品的详细,实现的效果如图 12-19 所示。

图 12-19　商品查询页面

## 4. 消息推送页面设计

消息推送页面是往 App 端推送消息,往数据库插入一条记录,所以只需要插入消息的内容即可,其他时间、是否使用等都可以使用默认的值。实现的效果如图 12-20 所示。

其他功能,如销量的查询、用户购买的商品查询、系统设置等功能,读者可以参考目前的功能进行实现。

图 12-20 消息推送页面

## 12.6 Spring Boot 服务端设计

视频讲解

服务器端的设计是最为重要的,主要负责数据的处理,即数据库的增加、删除、查询和修改操作。12.2 节中介绍 Spring Boot 知识的时候,已经了解了 Spring Boot 项目的完整结构,接下来针对"橙一色"App 项目中每个功能接口的设计进行详细讲解。

### 12.6.1 用户接口的设计

用户接口主要包含两个:用户登录以及用户注册。请求的方式都是 POST,接着需要去实现接口,根据登录或者注册是否成功返回结果值。Spring Boot 中是通过定义 Controller 来实现接口的调用的。首先在项目的 controller 包下定义一个叫作 OrangeUserController 的类,其具体内容如下所示:

```java
import com.OrangeSale.cn.entity.OrangeUser;
import com.OrangeSale.cn.service.UserService;
import org.springframework.beans.factory.annotation.Autowired;
import org.springframework.web.bind.annotation.*;
import java.util.HashMap;
import java.util.Map;
import java.util.Objects;
@RestController
@RequestMapping("/orange/user")
@CrossOrigin
public class OrangeUserController {
    @Autowired
    private UserService userService;
    @RequestMapping(value = "/login", method = RequestMethod.POST)
    public Map<String, Object> userLogin(@RequestBody OrangeUser orangeUser) {
        Map<String, Object> map = new HashMap<>();
        OrangeUser user = userService.userLogin(orangeUser);
        if (Objects.isNull(user)) {
            map.put("flag", false);
            map.put("msg", "login error");
            return map;
        }
        map.put("flag", true);
        map.put("data", user);
        return map;
```

```java
    }
    @RequestMapping(value = "/register", method = RequestMethod.POST)
    public Map<String, Object> userRegister(@RequestBody OrangeUser orangeUser) {
        Map<String, Object> map = new HashMap<>();
        Integer insertFlag = userService.userRegister(orangeUser);
        if (insertFlag > 0) {
            map.put("flag", true);
            map.put("msg", "register success");
            return map;
        }
        map.put("flag", false);
        map.put("msg", "register success");
        return map;
    }
}
```

以上即是一个 controller 的完整代码，向服务端发送请求的根地址是 http://localhost:8080/，但是每个项目在写 controller 的时候可以加上自定义的一部分地址。上述代码中第一部分加粗的代码：@RestController，这个注解说明只需要将方法中的返回值返回即可，不需要视图解析器，即不需要将返回值对应的页面返回，如 JSP 页面等；@RequestMapping("/orange/user")，在自定义类的上方加上这个注解，代表这个交互接口发送请求的地址是 http://localhost:8080/orange/user。

接着看用户登录部分的加粗代码：@RequestMapping(value = "/login", method = RequestMethod.POST)，这里 value 即是请求该接口的具体地址 http://localhost:8080/orange/user/login，method 是请求的方式，如 POST 或者 GET。下一部分的用户注册也同理，只需要将请求地址中的 login 换成 register 即可。

两个接口的逻辑是一样的，以登录接口为例讲解。登录接口的返回值是 Map<String, Object> 类型，方法的参数是 OrangeUser 实例对象，即跟数据库字段相对应的实体类，@RequestBody 是负责接口传过来的 JSON 字符串中的数据，调用 UserService 中的 userLogin() 方法，判断 userLogin() 方法的返回值是否为空、是否登录成功，然后将返回值返回。

下面介绍 UserService 中的 userLogin() 方法如何实现。在了解这个方法之前，先看一下 OrangeUser 这个实体类的内容，其他的实体类相同，这里只介绍一个，具体内容如下：

```java
public class OrangeUser {
    private Integer id;
    private String username;
    private String password;
    private String city;
    private String sex;
    //get()、set()方法
}
```

以上就是 OrangeUser 类的具体内容，下面是 UserService 类的内容：

```java
@Service
public class UserService {
    @Autowired
    private UserMapper userMapper;
    public OrangeUser userLogin(OrangeUser orangeUser) {
        return userMapper.userLogin(orangeUser);
    }
    public Integer userRegister(OrangeUser orangeUser) {
        return userMapper.userRegister(orangeUser);
    }
}
```

加粗代码: @Service, 这个注解写在类上, 代表将这个类交给 Spring 容器管理, Spring 容器要为它创建对象; @Autowired, 这个注解用于修饰变量, 写在变量上面, 并且由系统底层代理创建这个变量的实例, 并注入到这个类中, 就不需要手动去创建对象了。接着看一下 userLogin() 方法中的加粗代码: return userMapper. userLogin (orangeUser), 这句代码返回 UserMapper 中的 userLogin 中的返回值。UserMapper 是接口, 具体代码如下:

```java
import com.OrangeSale.cn.entity.OrangeUser;
import org.apache.ibatis.annotations.Mapper;
@Mapper
public interface UserMapper {
    OrangeUser userLogin(OrangeUser orangeUser);
    Integer userRegister(OrangeUser orangeUser);
}
```

为什么调用这个接口的方法就可以去做登录或者注册验证呢? 这其实是依赖于 MyBatis 的原理, 它将根据 UserMapper 这个接口和其对应的 mapper 文件去解析, 从而进行执行。UserMapper.xml 文件的内容如下:

```xml
<?xml version = "1.0" encoding = "utf-8"?>
<!DOCTYPE mapper PUBLIC "-//mybatis.org//DTD Mapper 3.0//EN"
        "http://mybatis.org/dtd/mybatis-3-mapper.dtd">
<mapper namespace = "com.OrangeSale.cn.mapper.UserMapper">
    <select id = "userLogin" resultType = "com.OrangeSale.cn.entity.OrangeUser">
    select * from orange_user where username = #{username} and password = #{password}
    </select>
    <insert id = "userRegister">
        insert into orange_user(username, password, sex, city) values(#{username},
                                                #{password}, #{sex}, #{city})
    </insert>
</mapper>
```

以上就是一个服务端请求的所有代码。当 Web 前端或者 App 端调用接口时, 执行的顺序即是: controller 中的方法接到请求调用 service 中的方法, 然后 service 中的方法是负责业务处理的, 它调用对应的 SQL 执行方法即 mapper 中的方法, 最终将返回值返回给前端进行页面渲染。至此,一个完整的请求就处理完成了。

使用 Spring Boot 与 MyBatis 整合时，需要在 application.properties 当中配置数据库的连接信息，以及 mapper 文件及对应的实体类的位置。具体内容如下所示：

```
server.port = 8081
spring.datasource.url = jdbc:mysql://localhost:3306/orange?useUnicode = true&characterEncoding = utf-8&useSSL = false
spring.datasource.driver-class-name = com.mysql.jdbc.Driver
spring.datasource.username = root
spring.datasource.password = 123456
mybatis.configuration.map-underscore-to-camel-case = true
mybatis.mapper-locations = classpath:mapper/*Mapper.xml
mybatis.type-aliases-package = com.test.vue.entity
```

### 12.6.2　商品接口的设计

12.6.1 节通过"用户接口的设计"对数据交互接口要用到的基本知识进行了较全面的讲解，后面交互接口涉及的相似内容就不再重复介绍，只对一些不同的地方做讲解。

商品的接口主要包含根据商品的 id 查询商品的详细信息，所有商品的信息查询，商品的更新、删除、添加，上传商品图片。OrangeProductController 类的具体内容如下：

```java
package com.OrangeSale.cn.controller;
import com.OrangeSale.cn.entity.OrangeProduct;
import com.OrangeSale.cn.service.ProductService;
import org.springframework.beans.factory.annotation.Autowired;
import org.springframework.beans.factory.annotation.Value;
import org.springframework.util.CollectionUtils;
import org.springframework.web.bind.annotation.*;
import org.springframework.web.multipart.MultipartFile;
import java.io.File;
import java.util.*;
@RestController
@RequestMapping("/orange/product")
@CrossOrigin
public class OrangeProductController {
    @Autowired
    private ProductService productService;
    @Value("${uploadFilePath}")
    private String uploadFilePath;
    @RequestMapping(value = "/searchById", method = RequestMethod.POST)
    public Map<String, Object> searchProductById(String productId) {
        OrangeProduct orangeProduct = productService.selectById(Integer.valueOf(productId));
        Map<String, Object> map = new HashMap<>();
        if (Objects.isNull(orangeProduct)) {
```

```java
            map.put("flag", false);
            map.put("msg", "search by id error");
            return map;
        }
        map.put("flag", true);
        map.put("data", orangeProduct);
        return map;
    }
    @RequestMapping(value = "/search", method = RequestMethod.GET)
    public Map<String, Object> searchProduct() {
        List<OrangeProduct> productList = productService.selectAll();
        Map<String, Object> map = new HashMap<>();
        if (CollectionUtils.isEmpty(productList)) {
            map.put("flag", false);
            map.put("msg", "search error");
            return map;
        }
        map.put("flag", true);
        map.put("data", productList);
        return map;
    }
    @RequestMapping(value = "/update", method = RequestMethod.POST)
    public Map<String, Object> updateProduct(@RequestBody OrangeProduct orangeProduct) {
        Integer updateFlag = productService.updateProduct(orangeProduct);
        Map<String, Object> map = new HashMap<>();
        if (updateFlag <= 0) {
            map.put("flag", false);
            map.put("msg", "search error");
            return map;
        }
        map.put("flag", true);
        map.put("msg", "update success");
        return map;
    }
    @RequestMapping(value = "/delete", method = RequestMethod.GET)
    public Map<String, Object> deleteProduct(@RequestBody OrangeProduct orangeProduct)
    {
        Integer deleteFlag = productService.deleteProduct(orangeProduct);
        Map<String, Object> map = new HashMap<>();
        if (deleteFlag <= 0) {
            map.put("flag", false);
            map.put("msg", "delete error");
            return map;
        }
        map.put("flag", true);
        map.put("msg", "delete success");
        return map;
    }
    @RequestMapping(value = "/add", method = RequestMethod.POST)
```

```java
public Map<String, Object> addProduct(@RequestBody OrangeProduct orangeProduct) {
    Integer addFlag = productService.addProduct(orangeProduct);
    Map<String, Object> map = new HashMap<>();
    if (addFlag <= 0) {
        map.put("flag", false);
        map.put("msg", "add error");
        return map;
    }
    map.put("flag", true);
    map.put("msg", "add success");
    return map;
}
@RequestMapping("/uploadCover")
public Map<String, Object> uploadCover(@RequestParam("file") MultipartFile file) {
    Map<String, Object> map = new HashMap<>();
    try {
        String fileName = file.getOriginalFilename();
        String changeFileName = uploadFilePath + fileName;
        File destFile = new File(changeFileName);
        if (!destFile.exists()) {
            destFile.getParentFile().mkdirs();
            file.transferTo(destFile);
        }
        map.put("flag", true);
        map.put("data", "http://localhost:8081/uploaded/" + fileName);
        return map;
    } catch (Exception e) {
        map.put("flag", false);
        map.put("msg", "upload error");
        return map;
    }
}
```

这里与用户接口不同的是增加了商品图片上传，首先看一下 uploadCover() 方法的参数：@RequestParam 用于将指定的请求参数赋值给方法中的形参，这里是指将上传的文件 file 赋值给 MultipartFile 实例对象，然后将文件上传到指定的地址：uploadFilePath，地址可以在 application.properties 中配置，如下所示：

```
#上传图片的路径
uploadFilePath=H:/uploaded/
```

然后将文件的访问地址返回给前端，这里要通过项目的地址访问图片，需要进行配置。配置类 WebMvcConfigure 的内容如下所示：

```java
import org.springframework.beans.factory.annotation.Value;
import org.springframework.context.annotation.Configuration;
```

```java
import org.springframework.web.servlet.config.annotation.*;
@Configuration
public class WebMvcConfigure extends WebMvcConfigurationSupport {
    private static final Logger logger = LoggerFactory.getLogger(WebMvcConfigure.class);
    @Value("${uploadFilePath}")
    private String uploadFilePath;
    @Override
    public void addResourceHandlers(ResourceHandlerRegistry registry) {
        logger.info(uploadFilePath);
        registry.addResourceHandler("/static/**").addResourceLocations("classpath:/static/");
        registry.addResourceHandler("/uploaded/**").addResourceLocations("file:" + uploadFilePath);
    }
}
```

上边加粗部分的代码是最主要的，其中 uploaded 即是访问图片的相对路径，使用项目的根路径加上相对路径再加上图片名字即可访问图片。如 http://localhost:8080/uploaded/a.jpg。uploadFilePath 即是图片上传地址。

接下来看一下 ProductService 的代码，与 UserService 相似，具体如下：

```java
@Service
public class ProductService {
    @Autowired
    private ProductMapper productMapper;
    public OrangeProduct selectById(Integer productId) {
        return productMapper.selectById(productId);
    }
    public List<OrangeProduct> selectAll() {
        return productMapper.selectAll();
    }
    public Integer updateProduct(OrangeProduct product) {
        return productMapper.updateProduct(product);
    }
    public Integer deleteProduct(OrangeProduct product) {
        return productMapper.deleteProduct(product);
    }
    public Integer addProduct(OrangeProduct orangeProduct) {
        return productMapper.addProduct(orangeProduct);
    }
}
```

按照用户接口设计的流程，接着就需要 ProductMapper 和其对应的 mapper 文件了。其中 ProductMapper 的具体内容如下所示：

```java
@Mapper
public interface ProductMapper {
    OrangeProduct selectById(@Param("productId") Integer productId);
```

```java
    List<OrangeProduct> selectAll();
    Integer updateProduct(OrangeProduct product);
    Integer deleteProduct(OrangeProduct product);
    Integer addProduct(OrangeProduct orangeProduct);
}
```

ProductMapper 接口对应着的 ProductMapper.xml 文件,主要包括了对数据库的增加、删除、查询和修改操作。具体内容如下所示:

```xml
<?xml version="1.0" encoding="utf-8"?>
<!DOCTYPE mapper PUBLIC "-//mybatis.org//DTD Mapper 3.0//EN"
        "http://mybatis.org/dtd/mybatis-3-mapper.dtd">
<mapper namespace="com.OrangeSale.cn.mapper.ProductMapper">
    <select id="selectById" resultType="com.OrangeSale.cn.entity.OrangeProduct">
        select id,name,img_url,price,add_time from orange_product where id = #{productId}
    </select>
    <select id="selectAll" resultType="com.OrangeSale.cn.entity.OrangeProduct">
        select id,name,img_url,price,add_time from orange_product
    </select>
    <update id="updateProduct">
        update orange_product set
        <trim suffixOverrides=",">
            name = #{name},img_url = #{imgUrl},price = #{price}
        </trim>
        where id = #{id} </update>
    <insert id="addProduct" parameterType="com.OrangeSale.cn.entity.OrangeProduct">
        insert into orange_product(name,img_url,price,add_time) values (#{name},
                        #{imgUrl},#{price},#{addTime})
    </insert>
    <delete id="deleteProduct">
        delete from orange_product where id = #{id}
    </delete>
</mapper>
```

以上加粗的代码即是对数据库 orange 中 orange_product 表的增加、删除、查询和修改操作。

### 12.6.3 消息接口的设计

通过前面两个接口的设计,读者已经了解了一个请求的完整处理流程,这里就不再详细列出每个文件的具体内容,只简单列出 controller 中的部分代码,读者可以在源代码的"橙一色"App 项目中看到完整的代码。

消息接口主要包含 3 个部分:消息的推送即插入数据库操作,消息的消费即读取记录,设置消息已读即更新操作。所以 controller 中包括 3 个方法,具体如下所示:

```java
@RestController
@RequestMapping("/orange/message")
```

```java
@CrossOrigin
public class OrangeMessageController {
    @Autowired
    private MessageService messageService;
    @RequestMapping(value = "/pushMessage", method = RequestMethod.POST)
    public Map<String, Object> insertMessage(@RequestBody OrangeMessage message) {
        Integer insertFlag = messageService.insertMessage(message);
        Map<String, Object> map = new HashMap<>();
        if (insertFlag > 0) {
            //参考前面 Map 键值对赋值,然后返回
        }
    }
    @RequestMapping(value = "/consumeMessage", method = RequestMethod.GET)
    public Map<String, Object> consumeMessage() {
        OrangeMessage orangeMessage = messageService.consumeMessage();
        Map<String, Object> map = new HashMap<>();
        if (Objects.isNull(orangeMessage)) {
        }
    }
    @RequestMapping(value = "/updateMessage", method = RequestMethod.GET)
    public Map<String, Object> updateMessage() {
        Integer updateFlag = messageService.updateMessage();
        Map<String, Object> map = new HashMap<>();
        if (updateFlag > 0) {
        }
    }
}
```

### 12.6.4 销量接口的设计

销量接口包含销量信息的查询、根据销量排序商品、销量信息的更新以及新的商品销量信息插入。根据销量信息排序是为了满足 App 端根据销量来查询商品的需求。销量类 OrangeSaleController 的具体内容如下所示：

```java
@RestController
@RequestMapping("/orange/sale")
@CrossOrigin
public class OrangeSaleController {
    @Autowired
    private SaleService saleService;
    @RequestMapping(value = "/getBySaleNum", method = RequestMethod.GET)
    public Map<String, Object> getBySalesNum() {
        List<OrangeProductSale> productSaleList = saleService.getProductBySalesNum();
        Map<String, Object> map = new HashMap<>();
```

```java
        if (CollectionUtils.isEmpty(productSaleList)) {
        }
    }
    @RequestMapping(value = "/searchSaleNum", method = RequestMethod.GET)
    public Map<String, Object> searchSaleNum() {
        List<OrangeProductSale> list = saleService.searchSaleNum();
        Map<String, Object> map = new HashMap<>();
        if (CollectionUtils.isEmpty(list)) {
        }
    }
    @RequestMapping(value = "/updateSaleNum", method = RequestMethod.POST)
    public Map<String, Object> updateSaleNum(@RequestBody OrangeSale OrangeSale) {
        Integer updateFlag = saleService.updateSaleNum(OrangeSale);
        Map<String, Object> map = new HashMap<>();
        if (updateFlag <= 0) {
        }
    }
}
```

以上代码即是销量接口的完整代码,查询操作一般使用 GET 方式的请求,而增加、删除、更新等操作多数情况下使用 POST 请求。

## 12.7 "橙一色"App 端功能实现

"橙一色"App 的注册登录数据存储是通过 Android 自带的 SQLite 数据库来实现的,商品的展示信息是通过在程序中模拟出来的,这种方式不利于数据的更新,这里采用 MySQL 数据库来实现,通过 Android 端发送请求,服务端返回验证数据的方式来操作。

### 12.7.1 App 登录/注册实现

视频讲解

要实现 App 的登录与注册,首先需要 App 端向服务端发送请求,然后根据请求结果判断是否登录及注册成功。

(1) 定义 OkHttpUser 请求类,用于实现向用户接口发送请求的功能,其中包含了用户登录验证以及用户注册的功能。具体内容如下所示:

```java
import android.util.Log;
import com.alibaba.fastjson.*;
import com.OrangeSale.cn.entity.OrangeUser;
import java.io.IOException;
import okhttp3.*;
public class OkHttpUser {
    public OrangeUser userLogin(OrangeUser orangeUser) throws IOException {
```

```java
        OkHttpClient okHttpClient = new OkHttpClient();
        MediaType mediaType = MediaType.parse("application/json; charset=utf-8");
        String jsonStr = JSONObject.toJSONString(orangeUser);
        RequestBody requestBody = RequestBody.create(jsonStr,mediaType);
        Request request = new Request.Builder()
                .url("http://172.16.40.55:8081/orange/user/login")
                .post(requestBody)
                .build();
        Response response = okHttpClient.newCall(request).execute();
        JSONObject jsonObject = JSON.parseObject(response.body().string());
        JSONObject jsonObject1 = jsonObject.getJSONObject("data");
        OrangeUser orangeUser1 = JSON.toJavaObject(jsonObject1,OrangeUser.class);
        return orangeUser1;
    }
    public boolean userRegister(OrangeUser orangeUser) throws IOException {
        OkHttpClient okHttpClient = new OkHttpClient();
        MediaType mediaType = MediaType.parse("application/json; charset=utf-8");
        String jsonStr = JSONObject.toJSONString(orangeUser);
        RequestBody requestBody = RequestBody.create(jsonStr,mediaType);
        Request request = new Request.Builder()
                .url("http://172.16.40.55:8081/orange/user/register")
                .post(requestBody)
                .build();
        Response response = okHttpClient.newCall(request).execute();
        JSONObject jsonObject = JSON.parseObject(response.body().string());
        Log.i("register","userRegister: " + jsonObject);
        return jsonObject.getBoolean("flag");
    }
}
```

（2）在 MainActivity 中调用登录验证的方法以及在注册 Activity 的 RegisterActivity 中调用注册方法。RegisterActivity 的代码如下所示：

```java
public class RegisterActivity extends AppCompatActivity implements View.OnClickListener, RadioGroup.OnCheckedChangeListener {
    private EditText usernameEdit, passwordEdit, surePasswordEdit;      //...组件初始化(省略)
    @Override
    protected void onCreate(@Nullable Bundle savedInstanceState) {
        super.onCreate(savedInstanceState);
        setContentView(R.layout.user_register);
        init();
    }
    @Override
    public void onClick(View v) {
        switch (v.getId()) {
            case R.id.reg_province:
                initCityPicker();
```

```java
                    cityPicker.show();
                    break;
                case R.id.reg_register:
                    //注册验证方法
                    //validateRegister();
                    Thread thread = new Thread(runnable);
                    thread.start();
                    break;
            }
        }
        //初始化城市选择器(同之前一样,此处省略)
        Runnable runnable = new Runnable() {
            @Override
            public void run() {
                //取值
                if (password.equals(surePassword)) {
                    if (!username.equals("") || !password.equals("")) {
                        if (!city.equals("")) {
                            ... //属性赋值
                            OkHttpUser okHttpUser = new OkHttpUser();
                            try {
                                if (okHttpUser.userRegister(orangeUser)) {
                                    Intent intent = new Intent(RegisterActivity.this, IndexActivity.class);
                                    intent.putExtras(bundle);
                                    startActivity(intent);
                                } else {
                                    Looper.prepare();
                                    Toast.makeText(RegisterActivity.this, "注册失败!", Toast.LENGTH_SHORT).show();
                                    Looper.loop();
                                }
                            } catch (IOException e) {
                                e.printStackTrace();
                            }
                        } else { ... //省略提示 }
                    }
                }
            }
        };
    }
```

以上 Runnable 部分的代码即是调用 OkHttpUser 中的 userRegister()方法,然后根据返回值进行验证是否登录成功。

（3）接下来验证注册这个接口的功能,首先注册一个新用户,看 orange_user 表中是否有新的用户记录。操作如图 12-21 所示。orange_user 表的数据如图 12-22 所示。

从图 12-22 即可看出,数据库新增了一个用户名为 xiaowang 的用户,说明注册功能已经实现了。登录功能的实现同注册实现类似,只是将调用的方法 userRegister()改为 userLogin(),然后对结果进行判断验证,这里不再赘述。

图 12-21 用户注册

| id | username | password | sex | city |
|----|----------|----------|-----|------|
| 1 | 123 | 123321 | 男 | 江西省 赣州市 章贡区 |
| 2 | xiaoming | 123456 | 女 | 江西省 赣州市 章贡区 |
| 3 | xiaoliang | 123456 | 女 | 江西省 赣州市 章贡区 |
| 4 | zsss | 123456 | 女 | 江西省 赣州市 章贡区 |
| 5 | xiaowang | 12345678 | 男 | 江西省 赣州市 章贡区 |

图 12-22  orange_user 表数据

## 12.7.2  App 首页商品信息展示功能实现

当登录或者注册成功以后,就要跳转到 App 的首页,首页中的热门商品数据之前都是通过程序模拟出来的,这里需要修改为从数据库读取出来的数据,然后解析出来渲染界面。具体操作如下。

(1) 定义 OkHttpClientProduct 类,用于请求商品信息接口,主要有商品信息的封装、商品图片的读取。具体内容如下所示:

```
public class OkHttpClientProduct {
    public List<OrangeProduct> getProduct() throws IOException {
        OkHttpClient okHttpClient = new OkHttpClient();
```

```java
        Request request = new Request.Builder()
                .url("http://172.16.40.55:8081/orange/product/search")
                .build();
        Response response = okHttpClient.newCall(request).execute();
        JSONObject jsonObject = JSON.parseObject(response.body().string());
        List<OrangeProduct> list = JSON.parseObject(jsonObject.getString("data"), new
                TypeReference<List<OrangeProduct>>() {
        });
        return list;
    }
    public Bitmap getImageBitMap(String imgUrl) throws IOException {
        Bitmap bitmap;
        OkHttpClient okHttpClient = new OkHttpClient();
        Request request = new Request.Builder()
                .url(imgUrl)
                .build();
        Response response = okHttpClient.newCall(request).execute();
        byte[] bytes = Objects.requireNonNull(response.body()).bytes();
        bitmap = BitmapFactory.decodeByteArray(bytes, 0, bytes.length);
        return bitmap;
    }
    public List<OrangeProductPack> getProductPack() throws IOException {
        List<OrangeProductPack> packList = new ArrayList<>();
        List<OrangeProduct> list = getProduct();
        for (OrangeProduct orangeProduct : list) {
            OrangeProductPack orangeProductPack = new OrangeProductPack();
            orangeProductPack.setId(orangeProduct.getId());
          orangeProductPack.setImgBitmap(getImageBitMap(orangeProduct.getImgUrl()));
            orangeProductPack.setAddTime(orangeProduct.getAddTime());
            orangeProductPack.setPrice(orangeProduct.getPrice());
            packList.add(orangeProductPack);
        }
        return packList;
    }
}
```

上述代码中 getProductPack() 方法负责将商品的信息进行封装返回给前端页面展示，之所以封装是因为直接读取的商品图片是一个 URL 地址，需要传递参数请求，将结果转换为 Bitmap 对象再进行渲染。使用 getImageBitMap() 方法即可实现请求图片的地址，然后将请求结果转换。

（2）IndexFragment 调用服务器端接口。IndexFragment 负责首页的展示，因为需要调用服务器端接口，网络请求耗时，所以需要使用一个异步任务来完成，否则在 Android 主线程中调用会报错。IndexFragment 与 12.6 节介绍的基本一致，只是将模拟数据的部分换成调用接口返回的数据。具体代码如下：

```java
public class IndexFragment extends Fragment implements View.OnClickListener {
    private SearchView searchView;
    private LinearLayout orangeLine, youziLine, juziLine, xiguaLine, liLine, appleLine,
            lemonLine, mangguoLine;
    private GridView gridView;
    private ProductAdapter productAdapter;
    private List<Product> productList;
    private List<OrangeProductPack> list = new ArrayList<>();
    @Override
    public View onCreateView(@NonNull LayoutInflater inflater, @Nullable ViewGroup
            container, @Nullable Bundle savedInstanceState) {
        View view = LayoutInflater.from(getActivity()).inflate(R.layout.content_index,
                container, false);
        init(view);
        return view;
    }
    private void init(View view) {                    //组件初始化
        new SearchProductTask().execute();
    }
    class SearchProductTask extends AsyncTask<Void, Void, List<OrangeProductPack>> {
        @Override
        protected List<OrangeProductPack> doInBackground(Void... voids) {
            OkHttpClientProduct clientProduct = new OkHttpClientProduct();
            try {
                list = clientProduct.getProductPack();
            } catch (IOException e) {     //...           }
            return list;
        }
        @Override
        protected void onPostExecute(List<OrangeProductPack> orangeProducts) {
            productAdapter = new ProductAdapter(getActivity(), list);
            gridView.setAdapter(productAdapter);
        }
    }
}
```

从上述代码可以看到，定义了一个 SearchProductTask 内部类，关于异步类在第 4 章已经介绍，不再赘述。这里 doInBackground() 负责处理网络请求，然后 onPostExecute() 在结果处理完成以后负责渲染界面。通过以上代码即可实现首页商品的展示，运行界面如图 12-23 所示。

（3）可以看到首页里显示的商品已经与之前的不一样了，接着查看一下数据库当中的商品，看是否与首页展示一致。orange 数据库的 orange_product 表数据如图 12-24 所示。

从图 12-24 可以看到，App 端显示的数据确实是从商品信息表里边读取的，说明 App 端的请求及服务器端的响应是正确的。

第12章 "橙一色"App 功能优化及完善

图 12-23 首页商品展示

| id | name | img_url | price | add_time |
|---|---|---|---|---|
| 1 | 苹果 | http://172.16.40.55:8081/uploaded/apple.png | 19.9 | 2019-12-16 15:58:05 |
| 2 | 香梨 | http://172.16.40.55:8081/uploaded/li.png | 9.99 | 2019-12-16 18:19:09 |
| 3 | 橙子 | http://172.16.40.55:8081/uploaded/orange.png | 222 | 2019-12-17 02:29:50 |
| 4 | 百香果 | http://172.16.40.55:8081/uploaded/baixiangguo.png | 3 | 2019-12-17 02:38:29 |

图 12-24 orange_product 表数据

### 12.7.3 App 商品信息展示功能实现

App 商品信息页面是针对商品页面功能的实现,其可以根据选择条件对商品的展示顺序进行调整。实现的原理主要是根据选择的条件去数据库当中查询,将带条件查询后的结果返回到 App 端进行界面渲染。本节不针对接口进行详细的介绍,读者可以参考之前的接口进行补充完整。ProductFragment 的具体内如下:

```java
public class ProductFragment extends Fragment {
    private Spinner conditonListSpinner;        //组件初始化(省略)
    @Override
    public View onCreateView(@NonNull LayoutInflater inflater, @Nullable ViewGroup
            container, @Nullable Bundle savedInstanceState) {
```

```java
        View view = LayoutInflater.from(getActivity()).inflate(R.layout.content_product,
                container, false);
        init(view);
        return view;
    }
    private void init(View view) {
        conditonListSpinner = view.findViewById(R.id.product_select_condition);
        initCondList();
        listViewAdapter = new ListViewAdapter(getActivity(), conditionList);
        conditonListSpinner.setAdapter(listViewAdapter);
        productGridView = view.findViewById(R.id.product_list);
        new SearchProductTask().execute();
    }
    private void initCondList() {
        conditionList = new ArrayList<>();
        Condition allCondition = new Condition();
        allCondition.setConditionIcon(R.drawable.all);
        allCondition.setConditionName("全部");
        Condition saleCondition = new Condition();
        saleCondition.setConditionIcon(R.drawable.salenum);
        saleCondition.setConditionName("按销量高低排序");
    }
    class SearchProductTask extends AsyncTask<Void, Void, List<OrangeProductPack>> {
        @Override
        protected List<OrangeProductPack> doInBackground(Void... voids) {
            OkHttpClientProduct clientProduct = new OkHttpClientProduct();
            try {
                orangeProductList = clientProduct.getProductPack();
            } catch (IOException e) {e.printStackTrace();}
            return orangeProductList;
        }
        @Override
        protected void onPostExecute(List<OrangeProductPack> orangeProducts) {
            ProductAdapter productAdapter = new ProductAdapter(getActivity(), orangeProductList);
            productGridView.setAdapter(productAdapter);
        }
    }
}
```

以上加粗部分的代码initCondList()方法即是对Spinner组件进行选择条件数据填充,可以为其设置监听事件,根据选择的条件去调用商品查询的接口,然后将结果返回去渲染界面。这里没有实现事件监听,只按照商品的添加顺序显示了商品,读者可以自行实现。

商品信息实现的最终效果如图12-25和图12-26所示。

图 12-25　商品界面

图 12-26　商品条件查询界面

## 12.7.4　App 消息推送功能实现

App 端的消息推送可利用 8.6 节设计的消息推送页面。要实现的功能包括：在消息推送页面上发送一条消息，在数据库中保存消息记录，然后 App 端收到消息并进行提示，提示完后消息设置为已读。下面讲解具体实现过程。

### 1. 消息推送页面推送消息

消息推送页面发送一条消息，也可以选择手动往数据库插入一条记录的方式，如图 12-27 所示。

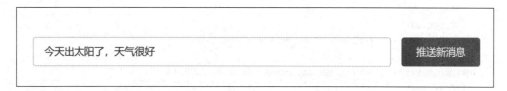

图 12-27　消息推送页面

### 2. orange_message 表记录

推送完消息以后，查看数据库中 orange_message 表的数据是否发生改变，如图 12-28 所示。

从图 12-27 和图 12-28 可以看出，消息已经推送完成，接着需要看 App 端是否有新的消息提示。

图 12-28　orange_message 表数据记录

### 3. App 端提示

App 端的消息提示功能采用的是 8.6 节介绍的 TimeService 类，之前是提示当前的时间，这里将其改为获取服务端的消息推送，然后提示。所以只需要将之前的 TimeService 进行部分改动即可。具体代码如下所示：

```java
public class TimeService extends Service {
    private static Timer timer = null;
    private NotificationManager manager;
    private NotificationCompat.Builder builder;
    @Nullable
    @Override
    public IBinder onBind(Intent intent) {
        return null;
    }
    @Override
    public int onStartCommand(Intent intent, int flags, int startId) {
        long period = 60 * 1000;              //1min 一个周期
        if (timer == null) {
            timer = new Timer();
        }
        timer.schedule(new TimerTask() {
            @Override
            public void run() {
                new ConsumeMessageTask().execute();
            }
        },0,period);
        return super.onStartCommand(intent,flags,startId);
    }
    class ConsumeMessageTask extends AsyncTask<Void,Void,Void> {
        @Override
        protected Void doInBackground(Void... voids) {
            /**
             * NotificationChannel 是 Android 8.0 新增的特性，如果 App 的
             * targetSDKVersion >= 26,
             * 没有设置 channel 通知渠道的话，就会导致通知无法展示.
             * 报错内容:Failed to post notification on channel "null"
             */
```

```java
            OkHttpMessage okHttpMessage = new OkHttpMessage();
            OrangeMessage orangeMessage;
            try {
                orangeMessage = okHttpMessage.consumeMessage();
                if (!Objects.isNull(orangeMessage)) {
                    Log.i("orangeMessage","run: " + orangeMessage.toString());
                    NotificationChannel channel = new NotificationChannel("channel",
                            "通知", NotificationManager.IMPORTANCE_DEFAULT);
                    manager = (NotificationManager) getSystemService(
                            Context.NOTIFICATION_SERVICE);
                    manager.createNotificationChannel(channel);
                    builder = new NotificationCompat.Builder(TimeService.this);
                    Date date = new Date();
                    SimpleDateFormat formatter = new SimpleDateFormat("yyyy-MM-dd HH:mm:ss");
                    builder.setContentTitle("来自"橙一色"的系统通知消息" + " " +
                            formatter.format(date));
                    builder.setSmallIcon(R.drawable.chengzi);
                    builder.setChannelId("channel");
                    builder.setContentText(orangeMessage.getContent());
                    builder.setAutoCancel(true);
                    builder.setDefaults(Notification.DEFAULT_ALL);
                    Notification notification = builder.build();
                    manager.notify(1,notification);
                }
            } catch (IOException e) {
                e.printStackTrace();
            }
            return null;
        }
    }
}
```

以上加粗部分的代码调用了 OkHttpMessage 类的 consumeMessage() 方法,该方法调用服务端的消息消费接口,并将消费后的消息设置为已读。该类的具体内容如下所示:

```java
public class OkHttpMessage {
    public OrangeMessage consumeMessage() throws IOException {
        OkHttpClient okHttpClient = new OkHttpClient();
        Request request = new Request.Builder()
                .url("http://172.16.40.55:8081/orange/message/consumeMessage")
                .build();
        Response response = okHttpClient.newCall(request).execute();
        JSONObject jsonObject = JSON.parseObject(response.body().string());
        OrangeMessage orangeMessage = JSON.toJavaObject(
                jsonObject.getJSONObject("data"),OrangeMessage.class);
        Request request1 = new Request.Builder()
                .url("http://172.16.40.55:8081/orange/message/updateMessage")
                .build();
```

```
            okHttpClient.newCall(request1).execute();
            return orangeMessage;
    }
}
```

consumeMessage()方法向服务端发送了两次请求:第一次是读取消息;第二次是设置为消息已读的请求。App端的提示如图12-29所示。

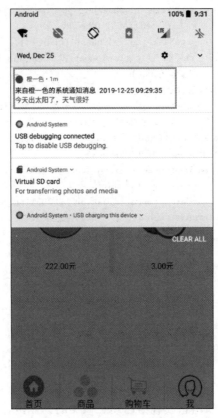

图 12-29  App 端消息提示

### 4. orange_message 表记录

接着再看一下 orange_message 表里的数据,查看刚刚推送的消息是否已经设置为已读。消费完以后的 orange_message 表的数据如图12-30所示。

| id | content | send_time | used |
|---|---|---|---|
| 1 | 测试新消息 | 2019-12-16 10:36:28 | YES |
| 2 | 添加一条新消息测试 | 2019-12-17 15:33:27 | YES |
| 3 | 测试退出消息 | 2019-12-17 15:34:32 | YES |
| 4 | 再次测试 | 2019-12-17 15:35:38 | YES |
| 5 | 今天天气真好 | 2019-12-17 15:39:20 | YES |
| 6 | 今天好热啊 | 2019-12-17 17:24:13 | YES |
| 7 | 今天出太阳了,天气很好 | 2019-12-25 17:29:28 | YES |

图 12-30  orange_message 表数据

从图12-30可以看到,App端消费完成以后,数据库中这条记录已经更新为已读的状态,说明整个消息推送功能的实现是正确的。

至此"橙一色"App的功能所用到的技术已经大体上讲解完成,剩下的功能读者可以根据已学到的知识进行完善。

## 12.8　本章小结

本章主要是利用Spring Boot+MyBatis+MySQL+Android技术完成了安卓端界面的设计以及服务器端数据处理的功能,同时也介绍了Spring Boot与MyBatis技术,在开发中,利用这些技术进行服务端的设计,可以节省大量开发时间,读者可以根据本章中的介绍在课下进行更加深入的学习。

# 参 考 文 献

[1] 汪杭军.Android 应用程序开发[M].北京:机械工业出版社,2014.
[2] 孙更新,吕婕,等.Java 毕业设计指南与项目实践[M].北京:科学出版社,2007.
[3] 杨丰盛.Android 应用开发揭秘[M].北京:机械工业出版社,2010.
[4] 王家林.大话企业级 Android 应用开发实践[M].北京:电子工业出版社,2011.
[5] 吴亚峰.Android 应用案例开发大全[M].北京:人民邮电出版社,2012.
[6] 韩迪.Android 创意实例详解[M].北京:北京邮电大学出版社,2012.
[7] 李刚.疯狂 Android 讲义[M].北京:电子工业出版社,2015.
[8] 极客学院.Android 课程体系讲解[EB/OL].https://www.jikexueyuan.com/.